VERY SHORT INTRODUCTIONS are for anyone wanting a stimulating and accessible way in to a new subject. They are written by experts, and have been published in more than 25 languages worldwide.

The series began in 1995, and now represents a wide variety of topics in history, philosophy, religion, science, and the humanities. The VSI Library now contains 300 volumes—a Very Short Introduction to everything from ancient Egypt and Indian philosophy to conceptual art and cosmology—and will continue to grow in a variety of disciplines.

Very Short Introductions available now:

ADVERTISING Winston Fletcher
AFRICAN HISTORY John Parker and Richard Rathbone
AGNOSTICISM Robin Le Poidevin
AMERICAN IMMIGRATION David A. Gerber
AMERICAN POLITICAL PARTIES AND ELECTIONS L. Sandy Maisel
THE AMERICAN PRESIDENCY Charles O. Jones
ANARCHISM Colin Ward
ANCIENT EGYPT Ian Shaw
ANCIENT GREECE Paul Cartledge
ANCIENT PHILOSOPHY Julia Annas
ANCIENT WARFARE Harry Sidebottom
ANGELS David Albert Jones
ANGLICANISM Mark Chapman
THE ANGLO-SAXON AGE John Blair
THE ANIMAL KINGDOM Peter Holland
ANIMAL RIGHTS David DeGrazia
ANTISEMITISM Steven Beller
THE APOCRYPHAL GOSPELS Paul Foster
ARCHAEOLOGY Paul Bahn
ARCHITECTURE Andrew Ballantyne
ARISTOCRACY William Doyle
ARISTOTLE Jonathan Barnes
ART HISTORY Dana Arnold
ART THEORY Cynthia Freeland
ATHEISM Julian Baggini
AUGUSTINE Henry Chadwick
AUTISM Uta Frith
BARTHES Jonathan Culler
BEAUTY Roger Scruton
BESTSELLERS John Sutherland
THE BIBLE John Riches
BIBLICAL ARCHAEOLOGY Eric H. Cline
BIOGRAPHY Hermione Lee
THE BLUES Elijah Wald
THE BOOK OF MORMON Terryl Givens
THE BRAIN Michael O'Shea
BRITISH POLITICS Anthony Wright
BUDDHA Michael Carrithers
BUDDHISM Damien Keown
BUDDHIST ETHICS Damien Keown
CANCER Nicholas James
CAPITALISM James Fulcher
CATHOLICISM Gerald O'Collins
THE CELL Terence Allen and Graham Cowling
THE CELTS Barry Cunliffe
CHAOS Leonard Smith
CHILDREN'S LITERATURE Kimberley Reynolds
CHOICE THEORY Michael Allingham
CHRISTIAN ART Beth Williamson
CHRISTIAN ETHICS D. Stephen Long
CHRISTIANITY Linda Woodhead
CITIZENSHIP Richard Bellamy
CLASSICAL MYTHOLOGY Helen Morales
CLASSICS Mary Beard and John Henderson
CLAUSEWITZ Michael Howard
THE COLD WAR Robert McMahon

COMMUNISM Leslie Holmes
THE COMPUTER Darrel Ince
CONSCIENCE Paul Strohm
CONSCIOUSNESS Susan Blackmore
CONTEMPORARY ART
Julian Stallabrass
CONTINENTAL PHILOSOPHY
Simon Critchley
COSMOLOGY Peter Coles
CRITICAL THEORY
Stephen Eric Bronner
THE CRUSADES Christopher Tyerman
CRYPTOGRAPHY
Fred Piper and Sean Murphy
DADA AND SURREALISM
David Hopkins
DARWIN Jonathan Howard
THE DEAD SEA SCROLLS Timothy Lim
DEMOCRACY Bernard Crick
DERRIDA Simon Glendinning
DESCARTES Tom Sorell
DESERTS Nick Middleton
DESIGN John Heskett
DEVELOPMENTAL BIOLOGY
Lewis Wolpert
DICTIONARIES Lynda Mugglestone
DINOSAURS David Norman
DIPLOMACY Joseph M. Siracusa
DOCUMENTARY FILM
Patricia Aufderheide
DREAMING J. Allan Hobson
DRUGS Leslie Iversen
DRUIDS Barry Cunliffe
EARLY MUSIC Thomas Forrest Kelly
THE EARTH Martin Redfern
ECONOMICS Partha Dasgupta
EGYPTIAN MYTH Geraldine Pinch
EIGHTEENTH-CENTURY BRITAIN
Paul Langford
THE ELEMENTS Philip Ball
EMOTION Dylan Evans
EMPIRE Stephen Howe
ENGELS Terrell Carver
ENGLISH LITERATURE Jonathan Bate
ENVIRONMENTAL ECONOMICS
Stephen Smith
EPIDEMIOLOGY Rodolfo Saracci
ETHICS Simon Blackburn
THE EUROPEAN UNION
John Pinder and Simon Usherwood
EVOLUTION
Brian and Deborah Charlesworth
EXISTENTIALISM Thomas Flynn
FASCISM Kevin Passmore
FASHION Rebecca Arnold
FEMINISM Margaret Walters
FILM MUSIC Kathryn Kalinak
THE FIRST WORLD WAR
Michael Howard
FOLK MUSIC Mark Slobin
FORENSIC PSYCHOLOGY
David Canter
FORENSIC SCIENCE Jim Fraser
FOSSILS Keith Thomson
FOUCAULT Gary Gutting
FREE SPEECH Nigel Warburton
FREE WILL Thomas Pink
FRENCH LITERATURE John D. Lyons
THE FRENCH REVOLUTION
William Doyle
FREUD Anthony Storr
FUNDAMENTALISM Malise Ruthven
GALAXIES John Gribbin
GALILEO Stillman Drake
GAME THEORY Ken Binmore
GANDHI Bhikhu Parekh
GENIUS Andrew Robinson
GEOGRAPHY
John Matthews and David Herbert
GEOPOLITICS Klaus Dodds
GERMAN LITERATURE Nicholas Boyle
GERMAN PHILOSOPHY
Andrew Bowie
GLOBAL CATASTROPHES Bill McGuire
GLOBAL ECONOMIC
HISTORY Robert C. Allen
GLOBAL WARMING Mark Maslin
GLOBALIZATION Manfred Steger
THE GREAT DEPRESSION
AND THE NEW DEAL
Eric Rauchway
HABERMAS James Gordon Finlayson
HEGEL Peter Singer
HEIDEGGER Michael Inwood
HERODOTUS Jennifer T. Roberts
HIEROGLYPHS Penelope Wilson
HINDUISM Kim Knott
HISTORY John H. Arnold
THE HISTORY OF ASTRONOMY
Michael Hoskin

THE HISTORY OF LIFE Michael Benton
THE HISTORY OF MEDICINE William Bynum
THE HISTORY OF TIME Leofranc Holford-Strevens
HIV/AIDS Alan Whiteside
HOBBES Richard Tuck
HUMAN EVOLUTION Bernard Wood
HUMAN RIGHTS Andrew Clapham
HUMANISM Stephen Law
HUME A. J. Ayer
IDEOLOGY Michael Freeden
INDIAN PHILOSOPHY Sue Hamilton
INFORMATION Luciano Floridi
INNOVATION Mark Dodgson and David Gann
INTELLIGENCE Ian J. Deary
INTERNATIONAL MIGRATION Khalid Koser
INTERNATIONAL RELATIONS Paul Wilkinson
ISLAM Malise Ruthven
ISLAMIC HISTORY Adam Silverstein
JESUS Richard Bauckham
JOURNALISM Ian Hargreaves
JUDAISM Norman Solomon
JUNG Anthony Stevens
KABBALAH Joseph Dan
KAFKA Ritchie Robertson
KANT Roger Scruton
KEYNES Robert Skidelsky
KIERKEGAARD Patrick Gardiner
THE KORAN Michael Cook
LANDSCAPES AND GEOMORPHOLOGY Andrew Goudie and Heather Viles
LATE ANTIQUITY Gillian Clark
LAW Raymond Wacks
THE LAWS OF THERMODYNAMICS Peter Atkins
LEADERSHIP Keith Grint
LINCOLN Allen C. Guelzo
LINGUISTICS Peter Matthews
LITERARY THEORY Jonathan Culler
LOCKE John Dunn
LOGIC Graham Priest
MACHIAVELLI Quentin Skinner
MADNESS Andrew Scull
THE MARQUIS DE SADE John Phillips
MARTIN LUTHER Scott H. Hendrix
MARX Peter Singer
MATHEMATICS Timothy Gowers
THE MEANING OF LIFE Terry Eagleton
MEDICAL ETHICS Tony Hope
MEDIEVAL BRITAIN John Gillingham and Ralph A. Griffiths
MEMORY Jonathan K. Foster
MICHAEL FARADAY Frank A. J. L. James
MODERN ART David Cottington
MODERN CHINA Rana Mitter
MODERN FRANCE Vanessa Schwartz
MODERN IRELAND Senia Pašeta
MODERN JAPAN Christopher Goto-Jones
MODERNISM Christopher Butler
MOLECULES Philip Ball
MORMONISM Richard Lyman Bushman
MUHAMMAD Jonathan A. C. Brown
MULTICULTURALISM Ali Rattansi
MUSIC Nicholas Cook
MYTH Robert A. Segal
NATIONALISM Steven Grosby
NELSON MANDELA Elleke Boehmer
NEOLIBERALISM Manfred Steger and Ravi Roy
THE NEW TESTAMENT Luke Timothy Johnson
THE NEW TESTAMENT AS LITERATURE Kyle Keefer
NEWTON Robert Iliffe
NIETZSCHE Michael Tanner
NINETEENTH-CENTURY BRITAIN Christopher Harvie and H. C. G. Matthew
THE NORMAN CONQUEST George Garnett
NORTH AMERICAN INDIANS Theda Perdue and Michael D. Green
NORTHERN IRELAND Marc Mulholland
NOTHING Frank Close
NUCLEAR POWER Maxwell Irvine
NUCLEAR WEAPONS Joseph M. Siracusa
NUMBERS Peter M. Higgins

THE OLD TESTAMENT
Michael D. Coogan
ORGANIZATIONS Mary Jo Hatch
PAGANISM Owen Davies
PARTICLE PHYSICS Frank Close
PAUL E. P. Sanders
PENTECOSTALISM William K. Kay
THE PERIODIC TABLE Eric R. Scerri
PHILOSOPHY Edward Craig
PHILOSOPHY OF LAW
Raymond Wacks
PHILOSOPHY OF SCIENCE
Samir Okasha
PHOTOGRAPHY Steve Edwards
PLANETS David A. Rothery
PLATO Julia Annas
POLITICAL PHILOSOPHY
David Miller
POLITICS Kenneth Minogue
POSTCOLONIALISM Robert Young
POSTMODERNISM Christopher Butler
POSTSTRUCTURALISM
Catherine Belsey
PREHISTORY Chris Gosden
PRESOCRATIC PHILOSOPHY
Catherine Osborne
PRIVACY Raymond Wacks
PROGRESSIVISM Walter Nugent
PROTESTANTISM Mark A. Noll
PSYCHIATRY Tom Burns
PSYCHOLOGY
Gillian Butler and Freda McManus
PURITANISM Francis J. Bremer
THE QUAKERS Pink Dandelion
QUANTUM THEORY
John Polkinghorne
RACISM Ali Rattansi
THE REAGAN REVOLUTION Gil Troy
REALITY Jan Westerhoff
THE REFORMATION Peter Marshall
RELATIVITY Russell Stannard
RELIGION IN AMERICA
Timothy Beal
THE RENAISSANCE Jerry Brotton
RENAISSANCE ART
Geraldine A. Johnson
RISK Baruch Fischhoff and John Kadvany
ROMAN BRITAIN Peter Salway
THE ROMAN EMPIRE
Christopher Kelly
ROMANTICISM Michael Ferber
ROUSSEAU Robert Wokler
RUSSELL A. C. Grayling
RUSSIAN LITERATURE Catriona Kelly
THE RUSSIAN REVOLUTION
S. A. Smith
SCHIZOPHRENIA
Chris Frith and Eve Johnstone
SCHOPENHAUER Christopher Janaway
SCIENCE AND RELIGION
Thomas Dixon
SCIENCE FICTION David Seed
THE SCIENTIFIC REVOLUTION
Lawrence M. Principe
SCOTLAND Rab Houston
SEXUALITY Véronique Mottier
SHAKESPEARE Germaine Greer
SIKHISM Eleanor Nesbitt
SOCIAL AND CULTURAL
ANTHROPOLOGY
John Monaghan and Peter Just
SOCIALISM Michael Newman
SOCIOLOGY Steve Bruce
SOCRATES C. C. W. Taylor
THE SOVIET UNION Stephen Lovell
THE SPANISH CIVIL WAR
Helen Graham
SPANISH LITERATURE Jo Labanyi
SPINOZA Roger Scruton
STATISTICS David J. Hand
STUART BRITAIN John Morrill
SUPERCONDUCTIVITY
Stephen Blundell
TERRORISM Charles Townshend
THEOLOGY David F. Ford
THOMAS AQUINAS Fergus Kerr
TOCQUEVILLE Harvey C. Mansfield
TRAGEDY Adrian Poole
THE TUDORS John Guy
TWENTIETH-CENTURY
BRITAIN Kenneth O. Morgan
THE UNITED NATIONS
Jussi M. Hanhimäki
THE U.S. CONGRESS Donald A. Ritchie
UTOPIANISM Lyman Tower Sargent
THE VIKINGS Julian Richards
VIRUSES Dorothy H. Crawford
WITCHCRAFT Malcolm Gaskill
WITTGENSTEIN A. C. Grayling
WORLD MUSIC Philip Bohlman

THE WORLD TRADE ORGANIZATION
Amrita Narlikar

WRITING AND SCRIPT
Andrew Robinson

Available soon:

COLONIAL LATIN AMERICAN LITERATURE Rolena Adorno

SLEEP
Steven W. Lockley and Russell G. Foster

FILM Michael Wood

MAGIC Owen Davies

ITALIAN LITERATURE
Peter Hainsworth and David Robey

For more information visit our website
www.oup.com/vsi/

Terence Allen and Graham Cowling

THE CELL

A Very Short Introduction

OXFORD
UNIVERSITY PRESS

Great Clarendon Street, Oxford OX2 6DP

Oxford University Press is a department of the University of Oxford.
It furthers the University's objective of excellence in research, scholarship,
and education by publishing worldwide in

Oxford New York

Auckland Cape Town Dar es Salaam Hong Kong Karachi
Kuala Lumpur Madrid Melbourne Mexico City Nairobi
New Delhi Shanghai Taipei Toronto

With offices in

Argentina Austria Brazil Chile Czech Republic France Greece
Guatemala Hungary Italy Japan Poland Portugal Singapore
South Korea Switzerland Thailand Turkey Ukraine Vietnam

Published in the United States
by Oxford University Press Inc., New York

First published 2011

British Library Cataloguing in Publication Data
Data available

Library of Congress Cataloging in Publication Data
Data available

Typeset by SPI Publisher Services, Pondicherry, India
Printed in Great Britain on acid-free paper by
Ashford Colour Press Ltd, Gosport, Hampshire

ISBN 978–0–19–957875–7

1 3 5 7 9 10 8 6 4 2

Contents

List of illustrations

Chapter 1
The nature of cells

What makes a cell?

A cell is the smallest unit of life. Everything living is formed of cells, from single-celled organisms, familiar to us as bacteria, to the most complex of creatures such as ourselves, formed of mind boggling numbers of cells, but trivial in comparison to cell numbers in two hundred tons of blue whale. In its role as the basic building block of life, a cell might be considered a relatively simple collection of components, gently 'ticking over' to maintain itself and occasionally dividing to create a new cell. Nothing could be further from the truth. Each and every cell, from the simplest to the most complicated, is a self-contained molecular factory working frantically throughout every minute of its lifespan, whether this is the half hour of unique existence of most bacteria before they divide, or the self maintenance and day-to-day activity of our nerve cells, living for several decades. The analogy of a cell as a factory falls somewhat short because, to match cellular activity, the factory itself and much of its machinery would have to be dismantled and rebuilt on a daily basis, without any slowing of production levels. Both animal and plant cells are around a thousand-fold larger than bacteria with a much more complicated and intricate internal organization.

Just what sort of chemistry can support the extreme levels of synthesis that allow the simpler cells to double themselves in minutes, and more complicated cells within a day? At the fundamental level, life is based on the atoms of only six of the 117 known elements: carbon, hydrogen, nitrogen, oxygen, phosphorus, and sulphur. Hydrogen and oxygen, combined as molecules of water, make up 99 out of every 100 molecules in the cell. This might appear to make life a rather dilute affair, but some of this water is tightly bound into the structure of larger molecules, and does not occur as actual liquid. Life at the molecular level is based on a restricted set of small carbon-based molecules common for all cells, which include sugars (providing chemical energy), fatty acids (forming cell membranes), amino acids (the units of all proteins), and nucleotides (the subunits of informational molecules such as RNA and DNA). All proteins are formed from just 20 different amino acids, which are common to every living thing. This 'alphabet' of amino acids is combined in a variety of different ways similar to the use of letters to make words, forming a massive 'vocabulary' of proteins. Proteins exist in a remarkably diverse variety of forms, providing the structural materials, chemical catalysts, and molecular motors that support and drive the processes of life. The code for each unique protein is stored in another code, this time of four letters, which makes up the genes in our DNA and which is passed from mother cell to daughter cell at each division. Each of the 24,000 or so individual genes in our DNA is specific for a single protein, but our bodies may have many times this number of proteins, produced by modifying the original genetic message. Proteins are combined to form multi-protein complexes, the cogwheels and bearings that drive the motors of production and maintenance within the cell. This level of complexity works perfectly for the simpler cells such as bacteria, but in larger and more complex cells such as our own, specific tasks are undertaken in separate sites in the cell termed organelles, which are separated from other components within the cell by their own membranes. Adding yet a further layer of complexity, our own bodies contain 200 or so different cell types.

This book attempts to provide an introduction to the massive diversity undertaken by cells in order to go about their business, and why any (cellular) shortcoming may result in disease.

Basic cell characteristics

Everything that lives on the surface of the planet is cellular in nature. At this point we should exclude viruses, as they are unable to reproduce themselves without hi-jacking the synthetic processes of the cell they infect. Their non-vital nature is emphasized by the capacity to make crystals of purified viruses in solution. The cell is the basic unit of life, and as such must fulfil three requirements: (1) to be a separate entity, requiring a surface membrane; (2) to interact with the surrounding environment to extract energy in some way for maintenance and growth; and (3) to replicate itself. These parameters are the same for all living beings, from the smallest bacterium, to any one of the 200 different cell types that create a human being. Many organisms live as single cells, whereas a human has some 100 trillion cells in all. This number can be compared with the total number of people on Earth today (6–8 billion), or even the total number of people estimated to have ever been on the planet (106 billion). As an aid for the perception of these extremely large numbers, we can perhaps use an analogy based on time. One trillion seconds ago equates to approximately thirty thousand years, a time when the Neanderthals were roaming around Europe.

A cell can function perfectly well as a single entity or, alternatively, one cell may be an infinitesimally small part of a massive community of cells that work together to make a single being such as ourselves. In multicellular organisms, groups of cells form tissues and tissues come together to form organs. Multicellularity requires cells with a complex internal architecture (as we shall see in Chapter 2), whereas the single-celled existence of a bacterium allows for a relatively simple organization (essentially a membranous bag containing the necessary chemical mix to maintain life). When life began, some four billion years ago, the first cells would have been similar to the bacteria of today. However, simplicity doesn't necessarily indicate primitive or unsuccessful, as bacteria are the most numerous and widespread of cells, and a branch of the bacterial family called Archaea can flourish in the most extreme environments on the planet, where nothing else can survive. In optimal conditions, some bacteria can reproduce every 20 minutes—a rate that will produce 5 billion bacteria in 11 hours, a number equivalent to the world's total human population. We ourselves are colonized by bacteria to the point where we house ten times the number of bacterial 'guests' (mainly in the gut, and weighing around one kilogram) than the number of actual cells in our own bodies.

Membranes and cell walls

As the basic unit of life, every cell must be a discrete entity and consequently requires its own boundary. This boundary is common to all life forms and consists of a thin membrane built from two layers of fat molecules (lipids) and coated and pierced with proteins that control the molecular traffic between the cell and its surroundings. Animal cells are usually combined to form tissues (for example, our skin), involving large numbers of different cell types. The membranes of these cells are in direct contact, held together at specific attachment sites, with other membrane areas modified to allow communication between adjacent cells. Unicellular organisms such as bacteria usually have an extra 'cell wall' outside of the membrane, often incorporating adhesive materials to 'glue' them to other cells or to surfaces (such as our teeth). Plant cells have a rigid cell wall woven from long molecules of cellulose. This major difference between plant and animal cell structure is largely the reason why animals move and plants (generally) do not.

Plant cell walls provide a strong mechanical framework as well as protection against pathogens and dehydration. Plant cell walls are

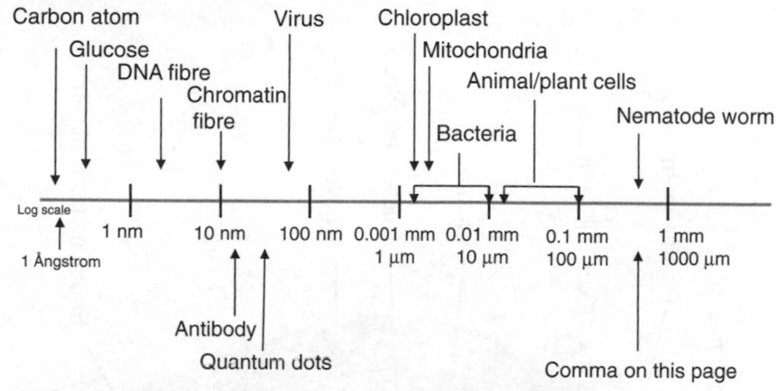

1. The sizes of atoms to relative simple worms on a log scale. Atoms are measured in Ångstroms, a 'dead' unit that is one tenth of a nanometre, but still in daily use by biophysicists

attached to each other by a glue made of pectin polysaccharides (the chemical that makes fruit preserves set), and further strengthened by the deposition of long strong molecules of celluloses and lignins that are the basic materials for the timber and paper industries. The rigidity of this type of construction allows massive and persistent growth (e.g. the giant sequoia trees of California) or longevity for thousands of years (e.g. bristlecone pines), but at the same time restricts plants to a rooted existence, although their leaves are well able to alter position to optimize exposure to sunlight.

The interior of the cell

In comparison to bacteria, plant and animal cells are massive, about one thousand times the volume. Figure 1 shows the scale of cells in the units they are measured in: nanometres (one millionth of a millimetre) cover the molecular sizes of cell components, and whole cells are usually tens of micrometres (one thousandth of a millimetre) in length. Plant and animal cells are also infinitely more complicated, containing a variety of structural elements built from proteins and several types of internal membrane-bound bodies called organelles (Figure 2). Individual organelles have

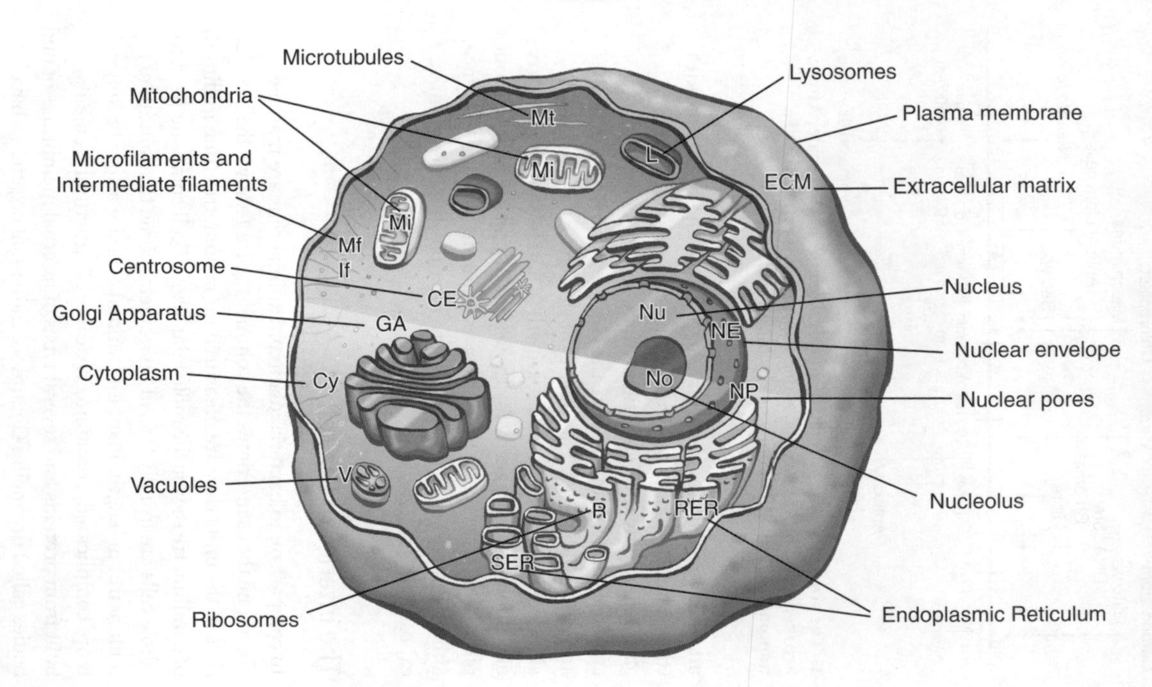
Microtubules
Mitochondria
Microfilaments and
Intermediate filaments
Centrosome
Golgi Apparatus
Cytoplasm
Vacuoles
Ribosomes
Lysosomes
Plasma membrane
ECM
Extracellular matrix
Nucleus
Nuclear envelope
Nuclear pores
Nucleolus
Endoplasmic Reticulum
Mt
Mi
Mf
If
CE
GA
Cy
V
R
SER
RER
L
Nu
NE
No
NP

2. Diagram of the contents of a cell

Centrosome (CE)—a pair of centrioles which organize microtubules according to the requirements of cell shape, movement, or division.

Cytoplasm (Cy)—all cell contents are suspended in a viscous fluid called cytosol.

Endoplasmic Reticulum (ER)—an extensive network of flattened membrane sheets. Rough endoplasmic reticulum (RER) has ribosomes for protein synthesis, smooth ER (SER) is involved in lipid metabolism.

Extracellular matrix (ECM)—material deposited outside the cell membrane, either as a thin layer, or larger amounts, such as collagen or bone.

Golgi Apparatus (GA)—a roughly circular stack of membranes which receives freshly synthesized proteins from the ER for modification, packaging, and distribution.

Lysosomes (L)—vacuoles containing lytic enzymes for breakdown of ingested material or cell debris.

Microfilaments and Intermediate filaments (Mf, If)—form the cytoskeleton, in combination with microtubules to bring about shape changes and cell movement.

Microtubules (Mt)—dynamic cytoskeletal components, which are constantly assembled and broken down to provide rigidity within the cytoplasm, and act as 'rails' for intracellular transport.

Mitochondria (Mi)—are the sites of energy generation for all cellular activity.

Nucleus (Nu)—contains the 'blueprint' for all cell activity, stored in code on the DNA, which necessitates constant and intense interaction with the cytoplasm.

Nuclear envelope (NE)—the double membrane separating nucleus and cytoplasm. The outer membrane is continuous with the endoplasmic reticulum.

Nuclear pores (NP)—thousands of channels in the nuclear envelope that control the rapid exchanges between nucleus and cytoplasm.

Nucleolus (No)—site of concentrated RNA and ribosome production.

Plasma membrane (Pm)—a lipid bilayer interspersed with proteins which encloses the cell, with specialized sites for attachment and communication with neighbouring cells.

Ribosomes (R)—numerous (millions per cell) molecular machines that assemble proteins.

Vacuoles and vesicles (V)—a variety of membrane-enclosed compartments, which fulfil specialized functions in particular cells

specific functions. Mitochondria, for example, produce the energy for all of the cell's activities from the breakdown of food molecules in animal cells. Plant cells uniquely have chloroplasts, which convert sunlight and CO_2 into sugars as an energy feedstock for their mitochondria. Both mitochondria and chloroplasts may themselves have been free-living organisms early in evolution, before becoming permanently incorporated into a larger, complex cell. Every cell has a 'blueprint' for its own creation coded by the DNA of its genes. In any particular organism, the DNA information content is the same in every cell type, whether they are brain, gut, or skin cells. Most cells (somatic cells) contain two copies of each DNA molecule (they are said to be diploid), except the germline cells (eggs and sperm) in which DNA exists as a single copy (these germline cells are haploid). When egg and sperm fuse to produce the first cell of the embryo (the zygote), two copies of the DNA are restored. How the DNA content of haploid sperm and egg cells is reduced will be discussed in Chapter 4.

In bacteria the DNA is circular, and lies naked within the cell contents, but in plant and animal cells it is contained within an organelle known as the nucleus (from Latin *nucula*, 'little nut'), where it is folded into chromosomes. Cells with nuclei are termed eukaryotes (a Greek word, meaning 'true kernel or nut'), whereas prokaryotes (from Greek, 'before nuts') such as bacteria have relatively little specialization of their contents into discrete internal organelles.

All cells reproduce themselves by splitting into two. Some bacteria manage to increase their contents rapidly enough to undergo division by a process known as binary fission in as little as 20 minutes. The much larger eukaryote cells may take the best part of a day to double their size before division. Cells as machines have an unparalleled efficiency of performance and diversity of components. The basic building blocks of cell are protein molecules, and every cell has tens of thousands of different proteins, in millions of copies. Actual numbers of the various molecules in a eukaryotic cell are extremely difficult to quantify, but estimates do exist for bacteria, which have 40% of their volume made up of around one million molecules of soluble proteins. A variety of five million small molecules account for 3%, then DNA 2%. The cell membrane and outer bacterial cell wall make up 20%, and the remainder of the contents are made up of the molecular machinery needed to synthesize proteins, including 2000 ribosomes. These figures can be roughly scaled up 1000-fold for the increase in volume of animal and plant cells, which may contain hundreds of individual organelles such as mitochondria, and around ten million ribosomes. Ribosomes are small molecular machines that provide for the assembly of new protein, a crucial requirement in cell maintenance and the provision of new protein prior to cell division.

It is hard to find man-made machinery that matches the workings of an average cell. Perhaps the biggest supercomputer on Earth would come close, but it would also need the ability to physically reproduce itself with faster and faster processors. While this may sound an extreme statement, it becomes more reasonable when you consider that cells have had around four billion years to get their act together, constantly driven by the unremitting pressures of natural selection. In simple terms, natural selection means that if a cell adapts to its environment, feeds and reproduces, then it survives, but failure means death. This process has produced a self-propagating, self-maintaining and repairing system that functions with a level of efficiency rarely approached by man-made machinery. Much of the entire subject of nanotechnology—the engineering of functional systems at a molecular scale—is aimed at replication of molecular reactions with the efficiency levels found in living cells. As well as their highly efficient metabolism (life-sustaining chemical activity), cells are also capable of producing unmatched structural rigidity, as found in the cells that give wood, palms, and bamboo values of mechanical performance that are exceptional in comparison to man-made equivalents (skyscrapers in the Far East often use local bamboo rather than steel scaffolding).

Tissues and differentiation

The main characteristic of eukaryotic cells is their ability to alter their shape, components, and metabolism to fulfil a particular task—to differentiate—a facility which allows them to come together and form multicellular tissues, to combine those tissues into organs, and then form an entire organism such as a human being. We are formed from around 200 different types of cells which make up the four main tissues: epithelia (surfaces), connective tissue (blood, bone, and cartilage), muscle, and nervous tissue. Cells are produced at widely different rates, from sperm (at a rate of 1000 in the same time as a heartbeat) to a nerve cell that may survive a lifetime. Some of our blood cells survive for only eight hours whereas the red cells circulate for about 120 days.

As an example of differentiation, we can briefly consider those cells that form our barrier to the environment. The cells at the surface of our skin have undergone a major 'remodelling' (differentiation) to become corneocytes, which are flattened polygonal plates made mainly of keratin (the same protein that makes nails, hair, and also feathers in birds). Each corneocyte spends about a day at the surface, before being shed and replaced by the one underneath, so that we present a new layer of skin cells to the world every day of our lives. The shed cells are replaced by division of unmodified cells that differentiate as they pass upwards through the 24 cell layers that make up the overall thickness of human skin. Shedding the top layer of corneocytes efficiently removes accumulated debris, along with the 7.5 million bacteria per square centimetre and the various fungal growths that constantly attempt to colonize our outer surfaces. There are 1000 corneocytes per square millimetre, and our overall skin surface area is just under two square metres, leading to a daily loss (and replacement) of around two thousand million cells each day.

The shed skin cells make up about 60% of household dust, and those we lose in bed provide the daily bread of the million or so dust mites that live in our mattresses. Not all parts of our bodies are replaced at this rate, but the skin, our largest organ, provides a good example of the basic properties of cells—replication, division, differentiation, time spent as a functional part of a tissue, and ultimately death.

Are single-cell organisms simple?

In stark contrast to the single purpose in life that a skin cell aspires to, we should consider what life is like for single-celled eukaryotic organisms. Because they are small, they could be considered 'simple' and 'primitive'. To survive, however, they need to seek out food, manage often hostile environments, reproduce, and avoid being eaten by other organisms. Protozoa are the largest group of singled-celled animals with *Amoeba* perhaps the most familiar. Amoebae have been observed to pursue and catch other protozoa such as *Paramecium*—perhaps the unicellular equivalent of a lion chasing a zebra. Possibly even craftier still are a group of protozoans called Suctoria. These unicellular organisms such as *Dendrocometes paradoxus* do not chase their food but instead attach themselves to various surfaces, extend their tentacles, and wait for some unfortunate protozoan, such as a *Paramecium*, to swim close by. If the tentacles are touched, the 'prey' is instantly paralysed, and the contents of its body are sucked down the tentacle into the body of the suctorian, reducing the prey to a shrivelled husk in a matter of minutes. The exact method for the transfer of prey contents is unknown, but it is driven by the spectacular arrays of cytoplasmic structures called microtubules inside the suctorian tentacles (see Figure 6c; microtubules will be described in Chapter 2). Having had a good meal, suctoria may choose to 'mate', a process requiring the attachment of modified tentacles to each other through which they exchange nuclei (they have one 'macronucleus' and three small ones). If bingeing and sex were not enough, *Dendrocometes* has another surprising facility. *Dendrocometes* lives attached to the gill plates of *Gammarus*, a freshwater shrimp. *Gammarus* moults regularly, so that *Dendrocometes* risks being left behind on an empty shell, losing the constant flow of water over the gill plates and the prey that this brings. However, *Dendrocometes* recognizes the earliest stages of moulting (possibly by responding to moulting hormone), and metamorphoses into a form bearing structures known as cilia which allows it to 'up sticks' and find a new set of gill plates (cilia will be described in Chapter 2). Thus, 'simple' single-celled organisms can have just as complex a lifestyle as that found in many multicellular organisms.

Although protozoa (and single-celled plants) must interact with their environment to survive, their perception is largely limited to physical and chemical interactions at their membranes, but with some interesting features. In single-celled plants such as the green algae *Chlamydomonas*, an 'eyespot' is visible within the chloroplast under a light microscope. The eyespot is a complex sandwich of membranes with rows of granules that contain around 200 different proteins, including the same rhodopsins found in the retina of our own eye. Signals from the photoreceptive eyespot cause the flagella (whip-like tails on the surface of the algae) to beat in different ways, so the algae swim towards brighter light, but away from light that is too bright. Whilst the eyespot might be thought of as a rudimentary eye, there is no imaging involved, or required, as the eyespot supplies all the information required for the organism's needs, helping it to cue day/night (circadian) rhythms, and optimize photosynthetic activity.

Tissue culture

An enormous amount of information about mammalian cells has come from *in vitro* studies; cells grown in a glass or plastic flask, maintained in a nutrient broth at 37 degrees in an atmosphere of

5% carbon dioxide, replicating the conditions in the body as closely as possible. Plant cells can also be cultured, often with their cell wall stripped off. Growth of cells and tissues outside the body goes back as far as the late 19th century, and the methodology for tissue and cell culture was established by Ross Harrison in Baltimore, USA around 1910. Tissues such as lung from laboratory mice embryos usually grow well as single cells, forming 'primary cultures', but they almost always have a limited life span, dying off after 50 or so rounds of cell division. Cells that are isolated from solid tissues require attachment to the surface of the plastic growth vessel to divide and grow, but blood cells, which normally exist in a liquid, do not require attachment, and grow as a 'suspension' culture. Cells are usually cultured as a single type, but mixed cell cultures such as those grown from isolated bone marrow will both survive and maintain the cellular interactions that occur in the body. When a small piece of solid tissue is excised and put into culture conditions, the cells that grow the most readily are those that respond to wounding in the living animal. These cells, known as fibroblasts, are the cells that maintain connective tissue that forms the structural framework of the body, such as ligaments and tendons. After injury, fibroblasts pull the edges of wounds together, and secrete the collagen that forms scar tissue. In culture conditions they are long and thin, with a leading edge that fans out at the front of the cell as it moves over the growing surface (Figure 3c). Within a few days, continuous rounds of cell division cause the surface of the flask to become increasingly crowded, and when there is no more room for newly divided cells to reattach to the growing surface, division will cease. This is called density-dependent inhibition of growth, and is characteristic of normal cells, but not those cells derived from tumours. At this point the cells are sub-cultured into new flasks at a reduced density. Because primary cultures die off after about 50 divisions, this puts a time limit on any series of experiments, so 'permanent' cell lines (usually derived from tumours) which divide indefinitely are often preferred by researchers.

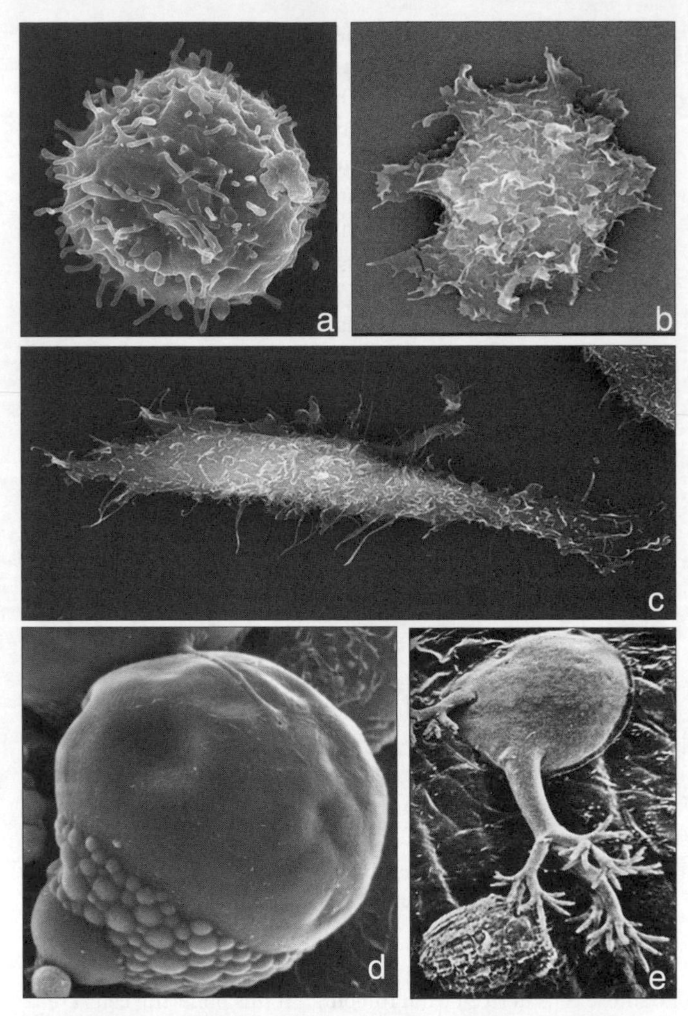

3. Cell types and their shape. (a) Haemopoietic (blood) stem cell (spherical), (b) epithelial (polygonal) cell, (c) fibroblast, (d) adipocyte or fat cell. All these cells have been grown in tissue culture. Although the surface of the fat cell is smooth, the other cells are more typical, with membrane folds, ruffles, and finger-like extensions (microvilli) over their surfaces. The difference in size between the fat cell and a normal cell is shown by the round cell at the bottom left. (e) *Dendrocometes paradoxus* feeding on a *Paramecium* attached to its tentacles

HeLa cells, the first permanent human cell line

Although permanent cell lines from humans are generally derived from tumours, tumour biopsies are notoriously tricky to establish in culture, with about 1 in 100 attempts successful. The first time that human cells were grown in culture as a permanent cell line was in 1951, by George Gey in Baltimore, USA. These were HeLa cells, so called because of their source—a cervical tumour biopsy from a lady called Henrietta Lacks. Henrietta's condition was diagnosed in Johns Hopkins, one of the very few hospitals in the USA in 1951 that would treat a member of the black population without health insurance. At the time, although cells from tumour biopsies worldwide were constantly put into culture conditions, none would survive longer than a few days. Henrietta's cells, however, began to double within days, sadly in line with the extremely aggressive and fast growing tumour spreading throughout her body, which killed her within months. The news of Henrietta's cells very quickly spread around the world of cell biology. In response to the worldwide demand for the first permanent human cell line, the Tuskegee Institute began mass producing the cells, shipping 20,000 tubes or 6 trillion cells every week. This means that every few months, the same volume of cells that had been enough to form Henrietta herself left the production line. HeLa cells were experimented on in every conceivable way, none of which would have been possible using 'whole' humans. They were exposed to every drug that might be toxic to tumours, every type of radiation possible, and a variety of toxins and viruses. For the most part they didn't react in a significantly different way to different species of mammalian cells, but it was important to show that human tissue culture cells did not behave in any unique ways. As her cells have now been grown worldwide for over half a century, there is probably enough of Henrietta (in the form of single cells) around the world for her to populate her own town. In fact one of her daughters, after seeing the film *Jurassic Park*, was briefly convinced that there were 'copies' of her mother in London, where she had read that much research had been carried out on HeLa cells. Henrietta's children would probably never have known about their mother's cells except that they were tracked down by Johns Hopkins researchers who wanted to compare the DNA from HeLa cells with its closest human match. This interest alerted them to the multimillion dollar industry that had been spawned by their mother's tumour, and, not unreasonably, they (and their lawyers) tried to extract some money from the situation. Unfortunately for them, in 1951 no rights to patient samples existed, and the state of California has subsequently ruled that 'a person's discarded tissues are not their property and cannot be commercialized'. Even now, Henrietta's children are still unable to afford health insurance in the USA. Despite 60 years of intensive research, including recent DNA analysis, the actual reasons for the prolific nature of HeLa cell growth are still unknown. HeLa cells could almost be considered the 'weeds' of tissue culture, having taken over many other cell lines by accidental cross contamination, a fact that only came to light when cultured cells began to be characterized by their DNA some 20 years ago.

For decades, research performed on cells *in vitro* has been questioned as to just how representative it is, because cells grow normally in a three-dimensional environment in a moving organism and are subject to a variety of stresses and factors not experienced by a single layer of cells growing in a plastic dish. Remarkably, in over half a century of concentrated research, cells in culture have produced very little in the way of misleading information, and without this accumulated knowledge it is unlikely that the current potential of stem cells for human therapy would exist at all.

Chapter 2
The structure of cells

All eukaryotic cells share the same basic layout, in that they are surrounded by a membrane, and filled with cytoplasm within which there is a variety of membrane-bound organelles that perform specialized tasks. One major organelle is the nucleus which contains DNA. At roughly one-thousandth the volume of the eukaryote cell volume, the organization of a prokaryote is relatively straightforward by comparison, although bacteria possess very similar structural aspects and machinery to eukaryotic cells. Their cell membrane is based on the same lipid bilayer (which we will discuss next) and their molecular machinery such as ribosomes for protein assembly functions in much the same way as eukaryotes. Bacteria do have specialized cell structures for motility such as flagella, and may possess internal membranes as in the case of a gas vacuole that serves as a buoyancy aid. However, the DNA in a bacterial cell is a single circular molecule and there is no separate nuclear compartment.

The cell membrane

All cells are enclosed by a boundary structure, the plasma membrane, which provides a barrier to other cells and the external environment. Although the membrane serves to contain the cell contents, unicellular organisms (bacteria included) usually have extra material on the outside and plant cells are characterized by a rigid cell wall of cellulose (Figure 4). Over a century of investigation has shown membranes to be an incredibly complicated and dynamic mixture of lipid (fat) molecules and proteins. Although the basic structure of the plasma membrane is extremely thin—just a couple of lipid molecules thick, forming a lipid bilayer—it is extremely tough and flexible, and also permeable to allow for the constant exchange of molecules between the cell and its surroundings. This is achieved via the water that constantly enters and exits in a controlled manner bringing soluble molecules like oxygen (needed as a fuel), and exporting waste products such as carbon dioxide. Large external material can be physically engulfed by the membrane, a process known as phagocytosis. The reverse process is exocytosis, in which a membrane-bound vacuole of material destined for export reaches the cell surface, at which point the membranes fuse and open to the outside, releasing the contents without breaching the overall integrity of the membrane. The dynamic nature of the cell membrane is such that the entire plasma membrane is 'turned over' (replaced) on an hourly basis.

Some of the earliest experiments involving the interactions of lipids and water to explain the properties of membranes were performed towards the end of the 19th century, on a kitchen table in Germany by Agnes Pockels. Agnes studied the behaviour of oil poured onto water in a flat dish, identifying the influence of impurities on the surface tension of fluids. She sent her results to Lord Rayleigh, who was sufficiently impressed to get them published in the scientific journal *Nature* in 1891. In 1932, Irving Langmuir, working in New York, won a Nobel Prize for showing that lipids spread on water produce a layer only one molecule thick and that all the molecules were orientated the same way. This happens because one end of the lipid molecule is attracted to water (it is hydrophilic), and the other end is repelled (it is hydrophobic). Each lipid molecule is shaped like an old-fashioned wooden clothes peg, with the top of the peg being the hydrophilic end, and the two legs of the peg representing the

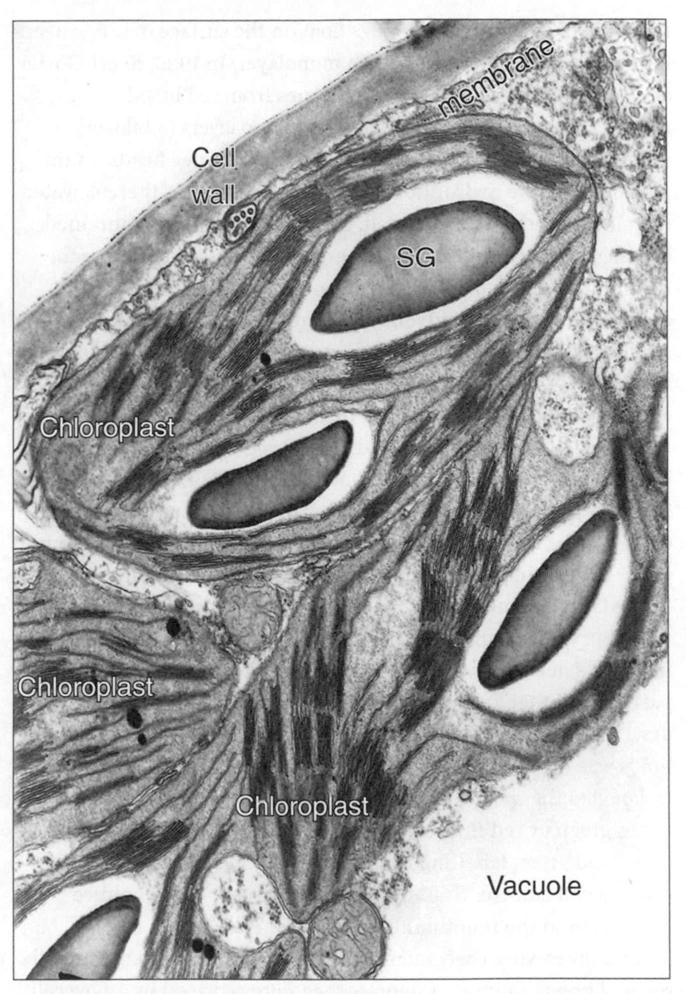

4. Section through a plant cell, showing the main differences from animal cells: a cell wall outside the cell membrane, chloroplasts with starch grains (SG) inside them, and a large vacuole in the centre of the cell

hydrophobic region. All the pegs float on the surface of water head down, legs uppermost, forming a monolayer. In 1925, Evert Gorter and James Grendel isolated membranes from red blood cells, finding that membrane was made up of two layers (a bilayer) of lipids, making a sandwich with the hydrophilic peg heads on the outsides, and the hydrophobic legs on the inside. As there is water both inside and outside the cell, this arrangement is maintained, keeping both hydrophobic surfaces together on the inside of the membrane bilayer. This arrangement was directly visualized decades later with the advent of electron microscopy, where thin sections at high magnification showed membranes as two dark lines separated by a light region between them (see, for example, Figures 4 and 14c). In 1935, James Danielli and Hugh Davson suggested that this lipid bilayer was covered on both sides by a layer of proteins, a model that lasted until 1972 when Seymour Singer and Garth Nicholson suggested that the proteins could also be threaded through the lipid layers, and project from either side of the membrane. Proteins with this configuration are termed transmembrane proteins, and a single protein might weave its way through the lipid bilayer several times. The lipid molecules in a membrane are highly mobile, constantly moving past one another and the membrane proteins, leading to the description of the 'fluid mosaic' membrane. This activity of the lipids was neatly demonstrated by Michael Edidin in 1970, when he labelled the membrane lipids of two different cell types with either green or red fluorescent chemicals. This gave a 'patchwork' of red and green labelling from individual cells grown together in a mixed cell culture. Edidin then added viruses to the culture which caused the membranes of adjacent cells to fuse to each other, thus mixing their membrane lipids and, within an hour, all red and green patches of fluorescence were replaced by an overall orange labelling, showing complete mixing of the individually labelled lipid molecules within the fused membranes.

Within this constant motion of the lipids, groups of membrane proteins float around freely in the lipid bilayer rather like ice floes in the polar seas. Sometimes a few lipid molecules will form a cluster for a few seconds creating specialized areas called membrane rafts. Membrane rafts were discovered only recently and their function is not yet completely understood, although it seems likely they are involved in signalling between cells. There are around 500 different membrane lipids which surround and anchor different proteins that can form channels through the membrane and control the continuous flow of molecules across it. These channels enable animal cells to maintain an internal concentration of sodium that is one-twentieth of the external concentration. This requires constant pumping to remove sodium which otherwise would raise the osmotic pressure in the cell which, in turn, would draw in water, with the potential to burst the cell. At the same time, potassium is maintained within the cell at a much higher concentration than the external levels, and the same membrane pump brings potassium in at the same rate as sodium is pumped out, an activity that takes about one-third of the total energy of the cell.

Protein molecules at the surface of the membrane also act as receptors for signalling molecules from outside the cell. These messages are then passed down a series of proteins within the cell to the nucleus to switch on genes, if required, in response to altered circumstances. Hormones such as insulin will interact directly with the membrane, allowing sugar to pass into the cell. Virtually everything that goes on in the cell either influences, or is influenced by, the activity of the membrane with its 500 types of lipid molecules and up to 10,000 types of membrane proteins. Cells 'feel' their immediate surroundings with fine extensions called microvilli (see Figure 3a–c). In some epithelial cells such as those responsible for nutrient uptake from the guts, the membrane becomes fashioned into a brush border, where tightly packed microvilli increase the surface area 30-fold (see Figure 14 in Chapter 5). While it is impossible to prioritize various parts of the cell, as they are mutually interdependent, without membranes, independent life could not exist.

Membranes inside cells

Membranes are also crucially important inside cells for two reasons: first, to provide surfaces on which chemical reactions can proceed, and secondly to provide separate areas inside the cell, allowing chemical reactions to proceed which might otherwise interfere with each other. In bacteria, the inner surface of the plasma membrane defines the position of everything within the cell and provides attachment points for intracellular contents that need to be in specific positions. Using the analogy of the cell as a factory, the internal membranes provide the workbenches, floors, ceilings, and walls for all the different parts of cell production, with the nucleus centrally positioned as the office in which the information is stored. In small cells, such as bacteria, which are usually rod shaped, the inside of the plasma membrane provides a large surface area in relation to the cell interior, so that anything that needs a fixed position can be 'hung' on the inside, and consequently bacteria and other prokaryotes generally have little or no internal membrane. As mentioned earlier, the internal volume of eukaryote cells is a thousand times that of a bacterium, so that the eukaryotic cell requires a vast internal membrane system around a hundred times the area of the plasma membrane itself. This internal segregation of biochemical activities is crucial as there are hundreds of chemical reactions going on that can seriously interfere with each other. Prokaryotes, with no internal membranes (and eukaryotes to some extent), get round this problem by aggregating groups of specific enzymes into multiprotein complexes, which work as free entities inside the cell. In addition, eukaryotes confine different metabolic processes within membrane-bounded compartments. The major internal membrane system in eukaryotic cells is the endoplasmic reticulum (usually shortened to ER), which forms a network throughout the entire cell. This part of the cell is called the cytoplasm (everything inside the cell membrane excluding the nucleus; Figure 5a, b) and everything in it is surrounded by the cytosol, a complex mixture of substances dissolved in water, like a very crowded 'molecular soup'.

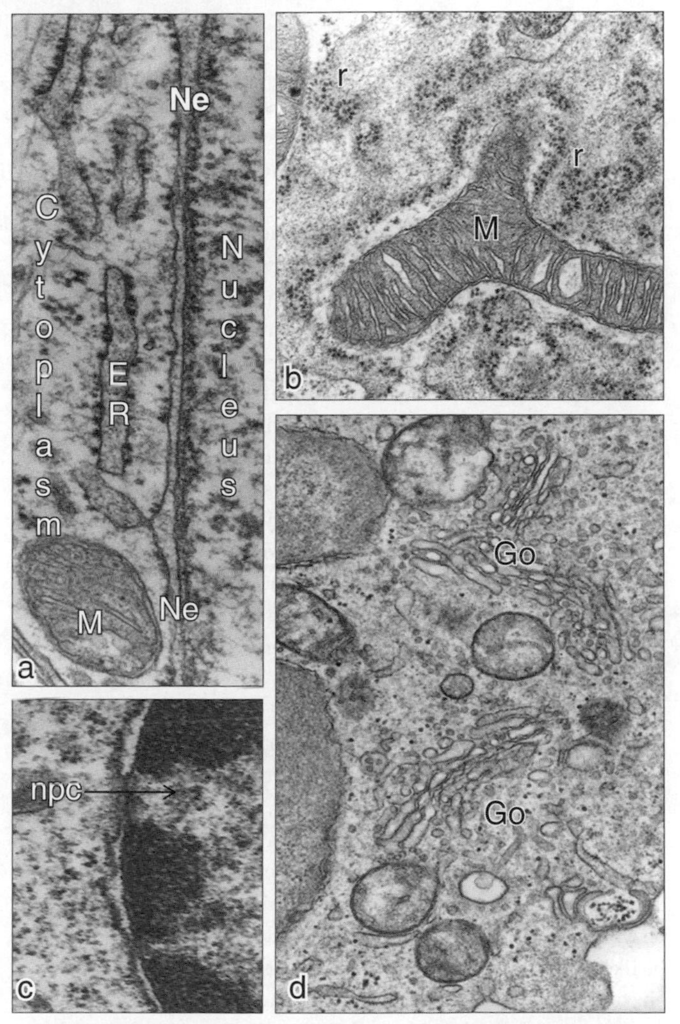

5. Internal cell membranes and structures. (a) The nuclear envelope (Ne) separates the nucleus from the cytoplasm, which contains other organelles including mitochondria (M) and endoplasmic reticulum (ER). (b) A mitochondrion surrounded by spiral polyribosomes (r) attached to the surface of the ER. (c) A nuclear pore complex, channel arrowed. (d) Golgi bodies (Go) comprised of stacks of membranes

Organelles

Two organelles in the cytoplasm—mitochondria and, in plants, chloroplasts—have double, rather than single membranes. This is most likely a hangover from when they were free-living forms early in cellular evolution. When they were incorporated into a larger cell, their own membrane became surrounded by the cell membrane of their host. Both mitochondria and chloroplasts contain DNA, further evidence that they were once free living. There are two theories as to how chloroplasts and mitochondria became part of eukaryotic cells: they could have 'invaded' the eukaryote cell or, alternatively, been engulfed by a larger cell, forming a relationship in which both partners benefit. The eukaryotic cell provided a 'safe' environment, in which the mitochondria generated energy that could be harvested by the host cell, and, in plant cells, chloroplasts produced glucose by photosynthesis. In mitochondria, energy is produced from glucose by a process called oxidative phosphorylation, which occurs on the surface of internal membranes called cristae (Figure 5b). In chloroplasts, glucose is produced by photosynthetic enzymes in stacks of membranes, called thylakoids (Figure 4).

All other membrane-bound organelles (collectively known as vacuoles or vesicles) have a single bilayer membrane. A typical cell will have around 1000 of these vacuoles, and a similar number of mitochondria. Secretory vesicles contain chemical messengers such as hormones for release from the cell. Endosomes, lysosomes, and peroxisomes (see Figure 2) all contain various mixtures of enzymes, proteins that catalyse chemical reactions. Lysosomes can be likened to the stomach of the cell, as they contain hydrolytic enzymes that break down biological material into its constituent parts to provide food for the cell. Lysosomes can also fuse with phagocytic vacuoles (phagocytes) containing engulfed material such as bacteria, killing and digesting the invading organisms. The Belgian scientist Christian de Duve discovered lysosomes, for which he received the Nobel Prize in 1974. He also discovered peroxisomes, which replicate by division like mitochondria and chloroplasts, but do not have their own DNA. Peroxisomes are involved in a variety of biochemical pathways and contain at least 50 different enzymes. They are important in the breakdown (oxidation) of substances such as fats, providing a major source of metabolic energy in animal, yeast, and plant cells. Because one of the products of oxidation is hydrogen peroxide, which is harmful to the cell, peroxisomes also contain an enzyme called catalase, which breaks down the hydrogen peroxide to water. Peroxisomes are also sites of synthesis of several enzymes, those in liver cells being responsible for the production of bile. As with most individual organelles, mutations in peroxisome formation have severe consequences, and any severe shortcoming will usually lead to a fertilized egg failing to develop past a few divisions.

The site of protein production for the contents of the various vacuoles such as lysosomes and peroxisomes is at the membranes of the ER. The ER was discovered by three pioneers of electron microscopy—Keith Porter and George Palade in New York and Fritiof Sjostrand in Sweden—in the early 1950s. Electron microscopy allows a 1000-fold increase in detail compared to conventional light microscopy, but imposes difficulties in the preparation of specimens in that an electron beam can only pass through extremely thin sections. In overcoming the difficulties of specimen preparation for electron microscopy, Porter and his colleagues opened up cell structure in a way that was previously unimaginable. At a stroke, indistinct and irregular shadowy lumps from light microscopy were viewed as sharply distinct organelles such as mitochondria (Figure 5b). In the words of Don Fawcett, a colleague of Porter and Palade, 'for morphologists, the decade from 1950 to 1960 held the same anticipation and excitement that attends the opening of a new continent for exploration'. The Fawcett atlases of biological ultrastructure of human tissues are still classics to this day.

Power to the cell

Mitochondria were first isolated biochemically and analysed by Alfred Lehninger in 1949, confirming the presence of the enzymes required for energy generation by oxidative phosphorylation, a highly efficient process in which nutrients are oxidized to produce adenosine triphosphate (ATP). The energy for doing work, building proteins and moving things around in cells is stored in a molecule of ATP. The energy in ATP is stored in 'high energy' phosphate bonds. To cut an involved biochemical process short, this energy comes from the release of electrons in the citric acid cycle within the mitochondrial membrane space, generating ATP synthase, an enzyme which then makes ATP. Energy is released from ATP when the phosphate bonds are hydrolysed (a process where the molecule is split into two parts by the addition of a molecule of water). With the release of energy, ATP is converted to ADP (adenosine diphosphate) which, in turn, is re-converted back to ATP, storing energy again, ready for the next release.

Only three years after Lehninger's biochemical characterization of mitochondria, Palade's electron micrographs showed their amazing membrane structure (Figure 5b). The order of these discoveries reflects an overall trend that the 'grind and find' workings of biochemistry have often generated seminal information on many cell parts in advance of their actual imaging in the electron microscope, although in the microscopist's view, nothing can compare with seeing what the constituents of the cell look like. The different approaches of the biochemist and biologist can be illustrated as follows. Presume neither had ever seen a wristwatch, but was presented with one for investigation. A few days later the biochemist would report that the watch had been analysed by separating it (grinding it up) into its component parts. This analysis would show that the watch was made from various proportions of copper, brass, steel, and bronze, with maybe a few diamonds. The biologist would hand the watch back intact, having done no more than maybe remove the back, and report that it seemed to house a spring which powered a series of interlocking cog wheels, that drove two arms on the front of the watch that seemed to rotate at a constant speed. While massively oversimplified, this comparison gives an idea of the differences between the analytical approach of the biochemist and the observational approach of the biologist. Fortunately, used together, these have been fruitful indeed in cell biology.

Protein production

Returning to the ER, the majority of ER membranes are covered with ribosomes (Figure 5a) and are known as rough ER, whilst the remainder bear no ribosomes (smooth ER). The job of ribosomes is to make (synthesize) proteins from amino acids, holding and joining the amino acids to make peptides, then polypeptides and complete proteins. A series of RNA molecules are involved in protein synthesis. Inside the nucleus, the sequence of nucleotide bases forming the code for a particular protein is first copied from the template DNA in a process called transcription, producing a new molecule of messenger RNA (mRNA). Messenger RNA then passes out of the nucleus, undergoing modification (called splicing) along the way. Once in the cytoplasm, ribosomes bind to messenger RNA, which then acts as a template for the linking together of amino acids into proteins, a process called translation. The amino acids are brought to the ribosome by short RNA molecules known as transfer RNA. Proteins made on the ER enter the space (lumen) between the ER membranes (Figure 5b), where they are folded into a final configuration before being passed on to other sites such as the Golgi bodies (Figure 5d). The Golgi body (or Golgi apparatus) is a stack of flattened membrane vesicles, where new proteins are packed into vacuoles for distribution throughout the cell, and may also have sugars added in a process known as glycosylation. Newly synthesized proteins undergo stringent quality control and, should they be defective in any way, they are tagged by molecules of ubiquitin for swift degradation. Protein misfolding is very detrimental, leading to disorders such as cystic fibrosis and diabetes. Protein quality control mechanisms may become less effective as we get older, leading to Alzheimer's disease and other age-related neurodegenerative conditions.

Once synthesized and folded, new proteins need to reach their final destination within the cell, amongst the other billions of protein molecules, constantly being synthesized and degraded. Some proteins may need to pass through one or two membrane barriers before reaching the site where they fulfil their function. In 1971, Günter Blobel and David Sabatini from the Rockerfeller Institute in New York suggested a 'signal hypothesis', in which proteins were given a luggage label, or zip code, to ensure they finished up in the right destination. Labelling takes the form of short sequences of amino acids, known as topogenic signals, which then attach to receptor proteins to allow them through membrane barriers to reach the correct destination. In 1999, Günter Blobel received the Nobel Prize for this work, which has explained the molecular mechanisms behind several diseases. Both cystic fibrosis and primary hyperoxaluria (a condition causing kidney stones at an early age) are caused by proteins failing to reach their correct destination. Blobel donated the million dollar prize money to the post-war reconstruction in Dresden, Germany, the country of his birth.

Lipid production

Besides its role in protein synthesis, the ER is a versatile organelle which can both receive and transmit signals and act as a cellular store for calcium, and is also responsible for the synthesis of lipids. Within individual cells, fat is produced at the surface of the ER as tiny individual droplets (lipogenesis). Although fat that we are familiar with around the edge of our steaks and often around our waistlines seems to be in solid homogenous lumps, it all exists within membrane-bound fat droplets in individual cells termed adipocytes (see Figure 3d). Given a continuous intake of nutrients, adipocytes will accumulate more and more lipid droplets, which coalesce with their neighbours to become larger and larger, accounting for the vast majority of the cell volume, which can reach over 100 times 'normal' size. Obesity is consequently a disorder of energy balance that results from the continued accumulation of lipid droplets within the adipocytes. At this point we might regret the efficiency of the ER, besides providing for fat storage, the ER also synthesizes enzymes on the smooth ER to break down fat by a process termed intracellular lipid hydrolysis or lipolysis. Thus, a major factor in body weight is the balance between the synthesis and breakdown of lipids in the ER. Considering the health consequences of being overweight, it is surprising that fat at the cellular level has received relatively little attention, with lipid droplets thought of as no more than simple storage depots. However, new studies are showing them to be remarkable organelles, and anything but 'lumps of fat'. All eukaryotic cells have the ability to make lipids, which produce all the naturally occurring oils and fats—from rapeseed and olive oil in plant cells to milk fats, lanolin, and lard in animal cells. Lipid molecules are concentrated at the surface of the ER, then pinch off forming a droplet (uniquely surrounded by a single lipid monolayer membrane) and remain adjacent to the ER, where the enzymes that catalyse lipid synthesis are located. Mitochondria are closely associated with the sites of lipid production, providing the energy for fat formation. These mitochondria are actually tethered to the surface of the ER by a group of membrane proteins. As more lipid is accumulated, individual droplets fuse with their neighbours, during which the separate membranes merge, sequentially producing ever larger droplets (Figure 3d). During fat breakdown, this process is reversed. Big droplets are fragmented into smaller ones, and enzymes from smooth ER break down the lipid molecules that protrude through the membrane, reducing the size of the droplet from the outside inwards.

Deposition of lipids can also occur within the cells found in the lining of blood vessels, particularly those forming the walls of major arteries. Here accumulation of lipids leads to formation of fatty plaques, resulting in atherosclerosis (hardening of the arteries), which limits blood flow and thus can lead to heart attacks and strokes. At other sites, interruption of blood flow can also produce kidney failure or gangrene. Excessive accumulation of lipids is also a major factor in type two diabetes and hepatic steatosis (fatty liver). Excess consumption of alcohol can cause changes in the way that the liver breaks down and stores fats, leading to more severe conditions such as cirrhosis. Fortunately, the fat droplets can still be broken down in the cell, so the condition is reversible with reduced consumption of alcohol. All this bad news leads to the reasonable question of whether we might not have been better off without fat cells, but they function in response to evolutionary pressures, allowing food storage that may well have helped us survive in times of shortage, and also allowing many other mammals to survive severe winters by hibernation.

Brown fat cells

There is another type of fat cell, termed brown fat cells, in which fat is broken down to generate heat by a process termed thermogenesis. In humans, babies have the most brown fat cells, usually in the shoulder areas. It was thought that brown fat was lost by adulthood in most humans, but in small mammals such as rats and mice, where heat loss is greater (from the increased surface area/volume ratio), brown fat cells are retained throughout life. Overfeeding mice and rats shows that they can burn off excess food intake as heat. Although hibernating animals build up massive stores of white fat to maintain themselves for months without food, their brown fat cells only get switched on at the time of waking to raise body temperature. When a technique called PET (positron emission tomography) scanning became a standard medical imaging technique a few years ago, some patients (wearing gowns only and therefore cold) showed mystery patches of high metabolic activity around the shoulders and back that disappeared in warm conditions. These patches were brown fat deposits, triggered into activity by the cold. Adults do in fact retain brown fat, and some individuals have significantly more than others. A likely reason why some individuals can eat as much as they like without gaining weight is that they have more brown fat. In theory, if we could switch our white fat cells into brown fat cells we could eat as much as we wanted with the end point of being hotter rather than fatter. Brown fat cells themselves differ from white fat cells by having many more mitochondria, with the iron in these mitochondria producing the brown colouration. The normal mitochondrial metabolism which generates energy stored as ATP is altered to produce heat by a process called proton leakage, produced by uncoupling proteins called thermogenins. Workers continuously exposed to cold conditions, such as deep sea divers, appear to accumulate much higher than normal amounts of brown fat deposits, indicating that brown fat can be regenerated in adults. If white fat cells could be converted to brown fat cells (as has been achieved in tissue culture), then we could have a useful tool in the battle against obesity, literally 'burning off' excess fat.

Modification of lipid production

In plant cells, genetic manipulation of the mechanism of formation of lipid droplets has staggering implications for seed crops. The biochemistry of production of oils in plant cells follows very similar routes to lipids in animal cells, and one group of enzymes (the diacyl glycerol transferases which catalyse triglyceride production) have already been genetically manipulated to approximately double the yield of oil and oleic acid in maize.

The cytoskeleton

A 'skeleton' instantly brings to mind the bony remnants of a long dead individual. The longevity of bone is due to the deposition of minerals such as calcium phosphate into the bone matrix by cells called osteoblasts, creating the structural rigidity. In life, the bones of the human skeleton are held together with tendons and ligaments, flexibly attached to each other as a system of levers, ready to generate movement which is driven by muscle contraction. There are structures with similar functions within cells, where microtubules play the role of levers and the 'muscle' activity is provided by actin microfilaments in association with myosin, which slide over each other to provide contraction much as they do in muscle itself. However, unlike the relative rigidity and permanence of our own musculo-skeletal system, the overriding characteristic of the cytoskeleton is one of extreme plasticity and dynamism, where components can be built (polymerized) from building blocks with remarkable speed and just as quickly removed by being broken down (depolymerization). The old-fashioned idea of a cell as essentially a balloon filled with jelly could not be more misleading; cell shape is controlled from within, modulated by signals received from the external environment, and capable of rapid response. Cells continually change their shape, change position relative to their neighbours, move through solid tissues, or take long journeys around the body by entering and exiting the bloodstream. Add to this the reorganization of the entire cytoskeleton required to separate chromosomes at division (as will be described in Chapter 4), and the dynamic nature of the cytoskeleton becomes its main characteristic.

An organized cytoskeleton is a property restricted to eukaryotes, although similar proteins do exist in a rudimentary form in some bacteria. The eukaryotic cytoskeleton is defined as a network of three types of large proteins: microtubules (formed from the smaller protein, tubulin); intermediate filaments (a group of fibrous proteins with similar properties); and microfilaments (formed from the smaller protein, actin) (Figure 6a, b). Each of these proteins has many associated proteins to help them fulfil their roles in just about every aspect of cellular function. Although each of the elements of the cytoskeleton provides specific parts of the overall function required to produce shape change and movement, the best way to think about the cytoskeleton is as an integrated system, involving all

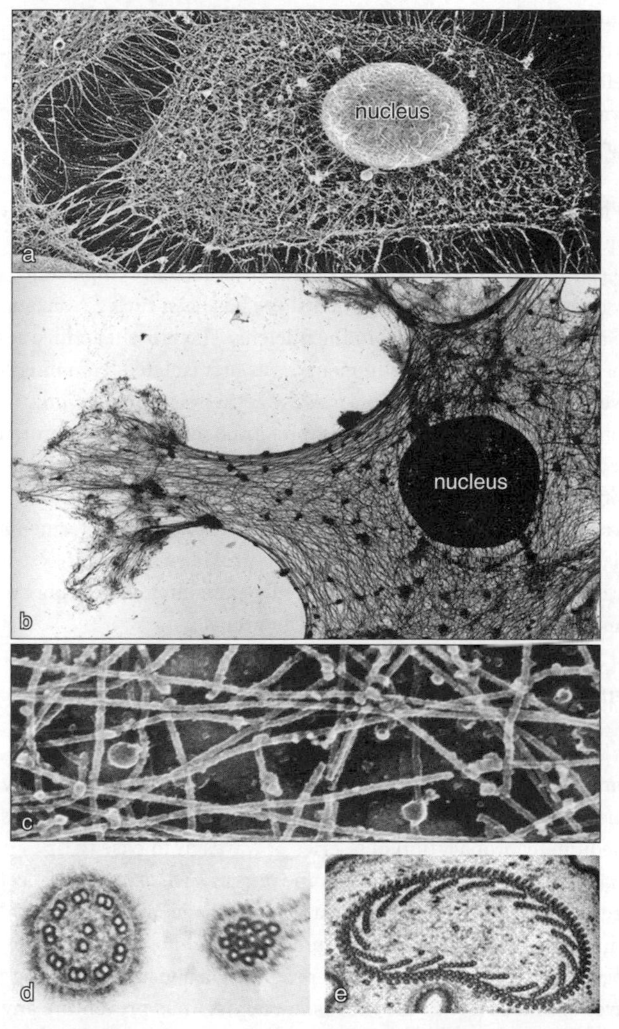

6. Cytoskeleton components. (a), (b) The network of fibres making up the cytoskeleton of intact cells, exposed by removal of cytoplasmic organelles leaving just the central nucleus. (c) Microtubules, that have been assembled in a test tube. (d) Section through a flagellum, showing the 9+2 arrangement of the axoneme. (e) Section through a suctorian tentacle, showing the microtubular arrays which surround the food canal

components together. As well as whole cell responses, the cytoskeleton also plays a crucial role in moving components within cells, where microtubules interact with motor proteins such as dynein, providing 'railway lines' for movement of vacuole-bound cargo or organelles throughout the cell.

Whether or not fibrous proteins permeate the nucleus to create an equivalent structure to the cytoskeleton has been controversial. Because the cytoskeleton is so obviously crucial to cytoplasmic organization, it is surprising that there has been such a resistance to an equivalent structure in the nucleus. The protein actin, one of the main cytoskeletal components, was first isolated from muscle over 70 years ago. Now, 'non-muscle' actin is accepted as a routine component of cytoplasm, and is in fact the most common protein in the cell. More recently still, actin has also become accepted as a constituent of the nucleus, forming part of a 'nucleoskeletal' arrangement of long filamentous proteins along with intermediate filaments of the nuclear lamina which provide a fibrous scaffolding for the arrangement of nuclear contents, forming the 'nucleoskeleton' (see Figure 7c in Chapter 3).

Cilia and flagella

Whip-like 'tails' have been observed on single cells from the earliest days of light microscopy in the late 17th century. Usually one or two such tails (flagella) move the cell through an aqueous medium by propagation of a series of waves from base to tip. In the epithelial tissues lining some organs such as the lungs, cells are covered in numerous flagella (called cilia in large numbers) which move a surface layer of mucus. In the windpipe, cilia beat together in a to and fro motion, producing a constantly moving layer of mucus upwards towards the larynx, thus preventing any accumulation of potential infective agents in the respiratory tract.

Some bacteria also possess flagella, but they are relatively simple, consisting of a rigid helical tail, which acts like a propeller, rotated at its base by a molecular motor. Eukaryotic cilia and flagella are rooted within the cell by a structure called a basal body, and generate their whip-like movement within the length of the flagellum itself by a system of microtubules, organized into a structure known as an axoneme. To quote Don Fawcett, in his classic 1961 work *The Cell*, 'few cellular activities have proved more fascinating to cytologists than ciliary and flagellar motion'. In 1887, Jensen squashed sperm flagella between a microscope slide and cover glass, describing how the sperm tails were 'frayed into a number of fibrils', some 60 years before this was confirmed by electron microscopy. Irene Manton, an English botanist who had managed to get an early electron microscope on the post-war 'Lend Lease' system from the USA, showed that there were 11 fibres in all plant flagella, matching those in animals, confirming that flagellar structure has been spectacularly conserved throughout evolution. The 'standard' axoneme structure consists of a central pair of microtubules surrounded by nine peripheral tubules (Figure 6d). Within the cell, the basal body is formed by a short cylinder of nine triplet microtubules without a central pair.

What makes a microtubule?

Each microtubule is a hollow tube with the wall made out of a protein called tubulin. Two molecules of tubulin form a dimer, which resembles a shelled peanut. These dimers join end to end making a long filament (protofilament), and 13 protofilaments are joined lengthwise to form the wall of the hollow tube that makes the microtubule (Figure 6c). The whole structure is stabilized by associated proteins. In the axoneme of flagella and cilia, movement is produced by a motor protein called dynein which links adjacent microtubules and allows them to slide over each other in a synchronized manner to produce a bend that travels down the flagellum, creating the 'whiplash' movement. We now know that the links between the dynein arms and adjacent tubules, which were discovered in the 1950s by Bjorn Afzelius, are successively made and broken, rather like climbing a rope hand over hand.

Should flagellar dynein be mutated or absent, then the consequences are significant. Twenty-five years after his initial discovery of the dynein links, Afzelius looked at the sperm of four patients at an infertility clinic in Sweden, and found that the dynein arms were absent in the sperm tail axonemes so that the sperm were 'non swimmers', which, not surprisingly, led to the infertility. Half of the patients also suffered from a condition known as *situs inversus*, where the major organs of the viscera (heart, spleen, and pancreas), which are normally on the left side of the body, become switched to the right. This turned out to result from a lack of functioning cilia early in embryo development, when the left–right body axis is established. This condition is called Kartagener's syndrome, after Manes Kartagener who described the condition in the 1930s.

A particular type of cilium is found in every cell. These 'primary cilia' act as sensory structures—rather like a radio aerial for collecting information from the immediate surroundings. They cannot move independently, as they lack both the central pair of microtubules and dynein links between the peripheral nine tubules. Primary cilia are now known to have a whole host of functions, acting as receptors for both mechanical and chemical stimuli. In the lining of the nose, modified primary cilia connect the receptors in the specialized cells of the olfactory epithelium in the nose (dendritic knobs) where smell is perceived. In the eye, the specialized photoreceptors of the retina are attached to their cell bodies by a primary cilium. The primary cilium also plays a controlling role in cell division, and is almost certainly involved in cell locomotion.

Diseases caused by defective cilia are known as ciliopathies, and they include a wide range of symptoms, often recognized and categorized as separate syndromes long before the underlying common cellular cause was identified. Some symptoms may be common to all patients, while others are unique. Patients with oral-facial-digital syndrome suffer from polydactyly (extra fingers and toes) and kidney problems. Patients with Bardet–Biedl syndrome (first identified in the late 19th century) also have kidney problems but in addition suffer from retinal degeneration which can lead to blindness, along with obesity and diabetes—all as a result of ineffective cilia.

Intracellular microtubules

Microtubules were thought to be limited to axonemes until improvements in electron microscopy preparation the early 1960s resulted in their discovery throughout the cytoplasm. Because they always appeared as straight rods, they were initially thought of as rigid and persistent structures. However, Lewis Tilney and Keith Porter showed that by merely cooling a protozoan called *Actinosphaerium* to around 4 degrees centrigrade, all the microtubule-supported cell extensions collapsed as the microtubules broke apart, subsequently re-forming after a few minutes at room temperature. Not until the 1980s did it become apparent just how dynamic microtubules actually were, when Tim Mitchison showed that microtubules could essentially collapse and re-form in seconds, a process that he termed 'dynamic instability'.

Microtubules also form the framework of the mitotic spindle, by which the chromosomes are distributed to daughter cells at division (see Chapter 4). By exposing dividing cells to a drug called colchicine (which binds to tubulin and stops it joining together to form filaments), the formation of the mitotic spindle can be inhibited, 'freezing' the process of division, and accumulating cells for chromosome analysis. Colchicine was the active ingredient in extracts from the autumn crocus *(Colchicum autumnale)* first used by the ancient Egyptians for arthritic conditions. Inhibition of mitotic spindle formation can also be achieved by drugs that stabilize cytoplasmic microtubules, preventing them from breaking down in order to re-form as spindle microtubules. One such drug is Taxol (extracted from the bark of the Pacific yew). Taxol became a potential blockbuster drug in cancer treatment, and because removal of the bark kills the tree, demand for the bark almost caused the loss of all Pacific yew trees in the USA. Fortunately, Taxol was subsequently chemically synthesized as paclitaxel. Due to the accelerated rate of division of cancer cells, almost any drugs that interfere with microtubules and spindle formation are potential cancer treatments.

Cultured fibroblasts have been the cells of choice for the study of microtubule function. Fibroblasts are found in connective tissue such as joints, ligaments, and tendons. Fibroblasts grown in culture are long and flattened, and move around the surface of the culture dish with a broad leading edge and a narrow trailing edge (Figure 3c). In contrast, cells from epithelia stay flattened and many-sided in culture (Figure 3b). In all cultured cells, cytoplasmic microtubules radiate outwards in the cytoplasm from a structure close to the nucleus called the centrosome. Centrosomes act as a microtubule organizing centre, controlling the turnover and distribution of microtubules. They contain structures (centrioles) identical to the basal bodies found at the base of each flagellum or cilium. These centrioles occur in pairs, positioned at right angles to each other. Early in cell division they separate and migrate to opposite ends of the cell to organize the microtubules that make up the mitotic spindle.

It is a short technical step from growing cells in plastic flasks to providing a suitable environment (a chamber at 37 degrees centigrade), allowing the flask to be placed on a microscope stage for living cells to be observed as they go about their business. Because living cells are largely transparent, it is hard to see much detail without various optical systems such as phase contrast microscopy, which convert small differences in the refractile properties in the cell components into light and dark regions. For this crucial advance, Frits Zernicke received the Nobel Prize in 1953. Nowadays, virtually any protein can be 'tagged' to fluoresce (when illuminated with UV light) by combining its genes with those of a protein called green fluorescent protein (GFP, proper name aequorin), originally extracted from a 'glow in the dark' jellyfish. By changing its sequence of amino acids, GFP has since had its fluorescence properties altered and is available in blue, orange, yellow, and red fluorescent varieties, allowing several different proteins to be followed over time in the same living cell. Add to this the ability of low-light cameras to capture signals from just a few molecules per cell, together with laser illumination and computerized imaging and analysis, and observing living cells currently provides a wealth of information that was unimaginable only a few years ago. Nowadays it is feasible to watch a particular living cell process over time in the light microscope, then 'flash freeze' the cell of interest in milliseconds and prepare it for examination by electron microscopy. Tiny probes called quantum dots, which are both fluorescent for light microscopy and electron dense for electron microscopy, allow labelling of the same molecules for both techniques.

An initial brief look at most living cells under the phase contrast microscope might be disappointing to the uninitiated, as not a great deal appears to be going on. Unicellular organisms will zoom around in real time, driven by their flagella or cilia, and amoebae crawl at a speed easy to see in real time. For cells in culture, the cellular activity seen in popular science programmes will invariably be time-lapse footage, with individual images recorded at intervals of a few seconds and then played speeded up. In this way, the hour a cell takes to divide is recorded with one image every ten seconds. Playing the images back at 25 frames per second compresses the process into just under ten seconds, making it appear much more dynamic.

Around the mid 1970s, time-lapse microscopy showed a movement inside cells which seemed unusual. As well as the continuous and random Brownian motion of the cytoplasm, there was a distinct stop/go movement, where a particle suddenly moved several micrometres across the cell, often stopping and moving on again. This 'saltatory' (related to leaping) movement surprisingly took place in straight lines, as though on rails. Saltatory movement was disrupted in the presence of cold or colchicine, conditions which break down microtubules, but not in the presence of paclitaxel, which stabilizes microtubules. Clearly, intact microtubules were acting like guide rails for vacuoles to travel along. Just how the material was driven along the microtubules was not discovered until the mid 1990s, when the motor proteins responsible for the movement were characterized. One such motor protein, kinesin, is shaped like an inverted 'Y' so that it has the molecular equivalent of two legs, and literally 'walks along' the microtubule, rather like a tightrope walker holding a large balloon (the attached vacuole) above his head. The cytoplasmic form of dynein (which drives flagellar microtubules past each other) works in a very similar way. The energy for both molecules is provided by the conversion of ATP to ADP. Each step is 16 nanometres, requiring 62 steps per micrometre of travel, and several micrometres (halfway across the cell) can be covered in a few minutes. The molecular interactions between kinesin and microtubules have been determined by resolving molecular detail in the electron microscope, helped by technology that permits instantaneous freezing of the cell, retaining molecular arrangements exactly as they were in life. Interactions of kinesin and microtubules are beautifully illustrated by molecular animations available on the Web (see Further reading).

Intermediate filaments

The requirements of a motile lifestyle in animals have resulted in the creation of mechanical strength by completely different ways. Whole organisms make a skeleton, either as a shell, as in the exoskeletons of insects and crustaceans, or an internal skeleton as in fish, amphibians, reptiles, birds, and mammals. In all cases, the skeletal material is made of proteins secreted by cells, and in some cases mineralized for further rigidity. This material is known as the extracellular matrix. Within individual animal cells, mechanical strength is provided by a group of proteins called intermediate filaments, high-tensile yet flexible cables which permeate the entire cell. Cells in tissues are joined to their neighbours by strengthened membrane junctions called desmosomes, which are anchored by intermediate filaments, providing a tensile network throughout the tissue.

Intermediate filaments are so called on account of their diameter (10 nanometres) which is intermediate between microfilaments (6 nanometres) and microtubules (25 nanometres). Although actin filaments and microtubules are the same in every cell (including plants), and intermediate filaments (lamins) in the nucleus are standard, cytoplasmic intermediate filaments are specialized according to their tissue type and embryonic origin. Cells in connective tissues are characterized by an intermediate filament called vimentin, whereas neurofilaments are found in nerve tissue, and desmin is characteristic of vertebrate muscle. Keratins are a large group of intermediate filaments found in epithelial cells, forming the structural protein of skin. Keratins secreted outside the cell form hair, wool, fingernails, horns, and hooves. These 'hard' keratins are stable extracellular secretions and consequently dead, although there are many other cytoplasmic keratins that are dynamic. Mutations in keratins can weaken skin, causing a condition known as epidermolysis bullosa, where even gentle friction can cause skin blistering so severe that it can be life threatening to new-born babies.

Microfilaments

The thinnest filamentous proteins in the cell are called microfilaments. They are around 6 nanometres in diameter, half that of intermediate filaments, and made from the protein actin. Actin is a globular protein (G-actin), but it takes a different form (F-actin) when assembled into filaments. Filamentous actin is then organized into bundles or networks by a host of actin-binding proteins, forming at least 15 different structures in non-muscle cells. The actin story goes back over 60 years, to the 1940s when Albert Szent-Gyorgi established the presence of both actin and myosin in striated muscle. Further work in the 1950s by Andrew Huxley and Hugh Huxley (unrelated) established that when muscle contracts, actin filaments slide over the myosin filaments, an interaction which shortens the muscle, producing force. The conversion of ATP to ADP releases the necessary energy for this molecular interaction to take place. Muscle has a highly organized geometrical molecular architecture, where every myosin molecule is surrounded by a 'cylinder' of six actin molecules allowing the molecules to slide over each other. This arrangement is only found in muscle cells, so for many years the possibility that actin and myosin could interact to produce contraction in non-muscle cells seemed unlikely. However, in 1973, Tom Pollard showed that there was more than one type of myosin in non-muscle cells. We now know that there are over 40 different myosins in mammals, and that (along with F-actin) they supply the motile force involved in cell division, cell movement, and the uptake of external material by cells (endocytosis). Actin also has a structural role in the cytoskeleton. Cores of actin filaments, bound together by the protein villin, support the finger-like projections of the cell membrane (filopodia and microvilli).

Cytoskeletal–nuclear interactions

From the interior of the nucleus to the surface of the cell, there are links between just about all of the filamentous proteins.
The lamins within the nucleus are a class of intermediate filaments (see Chapter 3), which are joined to the cytoplasmic intermediate filaments by protein bridges that cross the nuclear envelope. All the cytoskeletal elements are joined to each other, with direct protein links (plakins) between intermediate filaments and microtubules, and also between intermediate filaments and the microfilaments that form the third major element of the cytoskeleton. This interconnected protein scaffold of microtubules, intermediate filaments, and microfilaments, all with

different properties, functions together to maintain the structural and mechanical integrity of the animal cell and its ability to move (see Chapter 4).

Within a living cell, all three cytoskeletal components work in unison, as might be expected after four billion years of evolution. To explain the components of the cytoskeleton individually is like describing a piston, connecting rod, and crankshaft without mentioning a working engine. In both cases the whole is considerably more than the sum of the working parts.

Tensegrity

Thirty years ago, when Donald Ingber was an undergraduate at Yale University, he was convinced that the view of the cell as a 'rubber bag filled with jelly' was somewhat oversimplified. Ingber was intrigued by the revolutionary architecture of Buckminster Fuller in the 1940s, who created a series of robust structures called geodomes (including his own house). Geodomes are constructed from a shell of multiple small rigid triangles without any major supporting structures such as beams or columns. Fuller himself had been influenced by the sculptures of Kenneth Snelson, where rigid stainless steel rods appeared to float in thin air, but are actually supported by a system of cables, rather like the rigging on a sailboat, where the mast is kept in place by a balance of tension and compression. The mast itself is rigid in order to resist the compression produced by the tension in the rigging. The structure is robust, and will only fail if the mast buckles or the rigging breaks. This is the principle of tensile integrity or tensegrity, which offers the maximum amount of strength for the minimum expenditure of energy and materials. Ingber reasoned that tensegrity exists in every cell, mediated by rigid microtubules which resist the compression produced by actin and intermediate filaments. Tensegrity thus generates strength within all cell shapes, be they flattened hexagonal cells found in epithelia, or extended nerve axon cells which may be a metre in length. Tensegrity is at work even when a cell changes its shape, as happens in division. In culture, flattened or elongated cells will round up at the start of division, then pinch off leaving two spherical daughter cells, which then flatten and spread. As the edges flatten, triangles formed by actin fibres are clearly visible around the edge, with six neighbouring triangles forming a hexagon, exactly like the edge of a Buckminster Fuller geodome. The cell then flattens further, changing shape to a typical extended fibroblast form, forming attachments between the membrane and the base layer termed focal adhesions, before migrating as a single cell. In contrast, cells grown in culture from epithelial tissues will attach to their neighbours and move around as a sheet. As in living tissues, the epithelial cells attach to each other with structures called desmosomes, which are tough plaque-like structures formed by local reinforcement of the cell membrane, and anchored within the cell by intermediate filaments. In skin, a tissue constantly bent or stretched, epidermal cells have multiple desmosomes, and their intermediate filaments are strengthened by numerous keratin filaments (Figure 6a, b). When repeated through every cell, this arrangement creates an extremely tough tissue.

Chapter 3
The nucleus

Although the nucleus is the largest and most obvious organelle within the cell, the processes within it have proved rather more difficult to study than those in the surrounding cytoplasm. This may have been due to the difficulty of biochemical separation of its constituents. Largely as a result of improved technology over the last decade, we now know that the nucleus is the most spatially organized and probably the most dynamically active part of the entire cell. The production of DNA, RNA and the assembly of ribosomes involves a massive level of interaction with the cytoplasm and also constant repositioning of the nuclear contents. We shall consider nuclear structure from the outside inwards, beginning with the boundary between nucleus and cytoplasm—the nuclear envelope.

The nuclear envelope separates the nuclear contents from the cytoplasm, also controlling a constant and massive molecular interchange between the two compartments. So why do eukaryotes go to this trouble when prokaryotes like bacteria have no such partitioning yet reproduce themselves at such amazing rates? Successful as they are, at least in terms of numbers, bacteria can be considered essentially 'one trick ponies'. They have reached their limit as simple single cells, despite reproducing prolifically, and retaining sufficient genetic variability to consistently (and unfortunately) produce antibiotic-resistant strains. If the total number of organisms on the planet is equated with success, then bacteria come out on top. Conversely, in terms of biological complexity, they are also the most simple and consequently 'primitive' organisms on the planet. Bacteria also are the oldest at some four billion years, and as such provided the raw material for all subsequent life. The largest step in the evolution of living things on earth was the switch from prokaryotic to eukaryotic cell organization, i.e. the acquisition of a nucleus containing genetic material inside an isolating membrane. This has led to the proliferation and immense variation of life as we know it today. Just how the nucleus was acquired is uncertain, but it was probably as a result of phagocytosis of a small bacterium by a larger one. The smaller bacterium then 'took over' control of the larger one or endosymbiosis occurred with the partitioning of the DNA inside a membrane. Although cell biology is generally in agreement about the origin of mitochondria and chloroplasts by engulfment, no such consensus exists for the origin of the nucleus.

Enclosing the genetic blueprint of the cell within its own compartment has fostered the diversity we see in both unicellular and multicellular eukaryotic creatures. Each human produces around 150,000 different proteins, not in every cell, but across the various specialized tissues throughout the body. This is possible in spite of the fact that we only have 23,600 genes because the genetic message can be modified inside the nucleus after transcription (the transfer of information from DNA to RNA), and outside the nucleus (by the addition of simple chemical structures such as fats and sugars), thus increasing the overall number of possible protein products. For comparison, the simplest bacterium is probably *Mycoplasma genitalium* (found on primate genitalia) which has around 500 genes. The common gut bacterium *Escherichia coli* has 4,300 genes, whereas the smallest flu virus (which needs to hijack the machinery of the cells it infects to reproduce itself) has but 11 genes.

Separation of nuclear contents and cytoplasm has resulted in eukaryotic cells becoming much larger and more complex in comparison to prokaryotes. The circular molecule of DNA in bacteria is tacked on to the inside of the cell membrane at various points and may stretch around the whole cell. This is fine for a DNA genome with 4.6 million nucleotide base pairs (nucleotide sequences carry the genetic code) as is the case for *E. coli*. With a 1000-fold increase in length for the DNA in human cells however, accessing a specific gene of around a few hundred base pairs from the overall 3.1 billion base pairs has to be easier when the DNA is concentrated within the nucleus. The restriction of DNA to the nucleus in eukaryote cells also avoids any potential interference with the sophisticated workings of the cytoskeleton and cytoplasmic organelles. There are no such problems in prokaryotes, where the DNA is short (and circular), and there is little if any cytoskeletal structure.

The nuclear envelope and pore complexes

The nuclear envelope consists of two distinct membranes, the outermost being formed by endoplasmic reticulum, which is separated from the inner nuclear membrane by a perinuclear space (Figure 5a). The inner nuclear membrane is lined with a network of fibrous proteins which form a structure known as the nuclear lamina. Both the nuclear membrane and the lamina below it are pierced by nuclear pore complexes, which control the flow of everything into and out of the nucleus, apart from very small molecules which can pass directly through the nuclear envelope. There are around 5,000 pore complexes distributed over the surface of the nucleus in mammalian cells. Nuclear pore complexes are made from 50 proteins (nucleoporins), and are the largest molecular machines in the cell. Nuclear pores connect the inner and outer nuclear membranes, and also project eight cable-like proteins into the cytoplasm, and eight further fibres into the nucleus, forming a structure rather like a basket (Figure 7d, e). Inbound cargoes of molecules attach to the fibres extending outwards from the pore, and are then passed through the membrane channel and out of the basket into the nucleus.

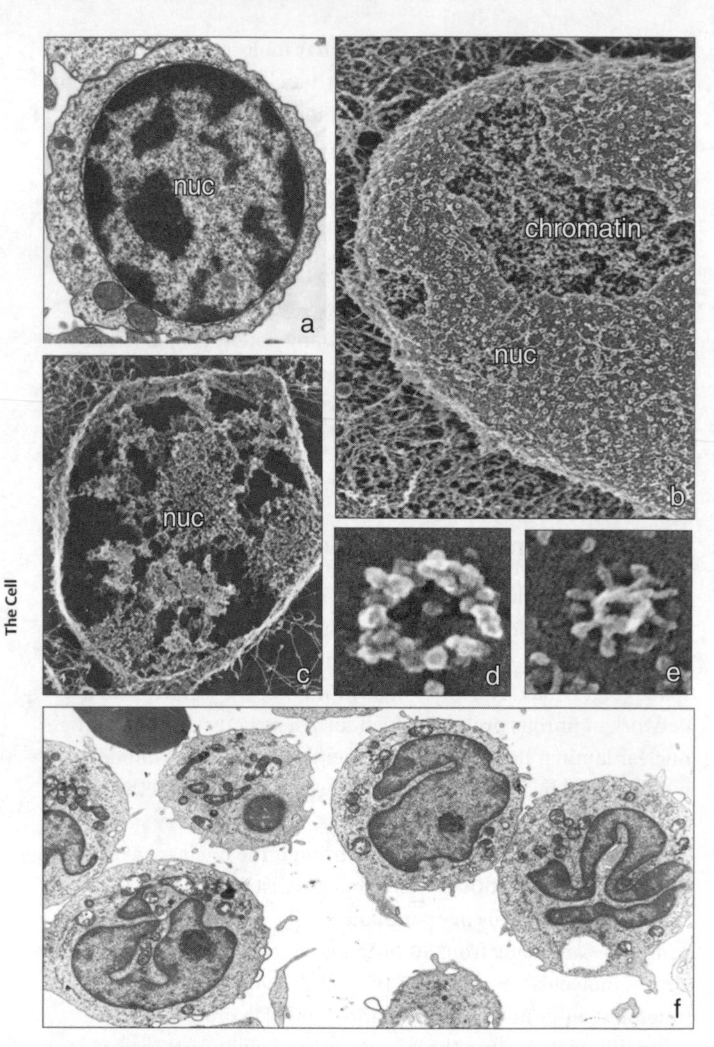

7. The nucleus. (a) Section through an intact nucleus, (b) surface view, with the internal chromatin exposed by removal of part of the pore covered nuclear envelope, and (c) the nucleolus (nuc) and nucleoskeleton after DNA removal, (d), (e) nuclear pores viewed from the outside and inside of the nucleus, (f) leukaemia cells containing distorted nuclei

A scaled-up analogy of pore traffic activity would involve a short length of drainpipe (as the pore channel) through which a mixture of tennis balls, golf balls, and marbles would pass in both directions at a rate of 1000 journeys per second. The flow is controlled by nucleoporin proteins which project into the channel, sorting and propelling the various molecules in the correct direction.

Each protein cargo is 'tagged' by an amino acid sequence that acts like a luggage label to ensure that they finish on the correct side of the nuclear membrane. The actual passage through the pore requires attachment of 'chaperone' proteins called importins or exportins, which accompany the cargo through the pore but are then chopped off as the cargo exits the pore and reattached to more cargo. Ribosomes are assembled from RNA (made in the nucleus) and proteins (made in the cytoplasm), and consequently generate a high level of pore traffic regardless of other nuclear/cytoplasmic exchange. In a HeLa cell, ten million ribosomes are produced each day. Seven thousand are produced each minute, each having around 80 proteins, requiring the production of half a million proteins per minute in the cytoplasm. These proteins are imported into the nucleus at a rate of 100 per pore per minute, passing (amongst other traffic) three ribosomal subunits on their way out of the nucleus. Certain diseases are directly associated with nucleoporin proteins. In primary biliary cirrhosis, proteins (autoimmune antibodies) are produced that attack nucleoporins, eventually leading to complete cirrhosis of the liver.

Although the nuclear envelope clearly separates nucleus and cytoplasm, it also physically links them. Proteins called nesprins, which are anchored in the inner nuclear membrane, reach across the perinuclear space, pass through the outer nuclear membrane, and extend for some distance into the cytoplasm, where they attach to the cytoskeleton. Nesprins are some of the largest proteins in the cell. This attachment of molecules from within the inside of the nucleus to cytoskeletal elements (which themselves are linked to the plasma membrane) means that there is a potential molecular linkage directly from the cell surface through to the nucleus, an interesting but as yet unexplained linkage.

The nuclear lamina

The nuclear lamina was originally visualized in the electron microscope as a fibrous matrix on the inside of the inner nuclear membrane. These protein filaments resist stretching and form the 'high-tensile cables', closely related to the intermediate filaments of the cytoskeleton. Thus, the nuclear lamina protects the nuclear contents from mechanical stress, and also anchors the position of the nucleus in the cell, providing sites for attachment to the cytoskeleton in the cytoplasm. A structure called the centrosome, which is the main microtubule organizing centre of the cell, is also kept close to the nuclear surface by attachments to the nuclear lamina. Besides these mechanical functions, the nuclear lamina also plays a major role in the overall organization of nuclear contents affecting both gene regulation and the passage of genetic information to the cytoplasm. Gene defects that lead to disruption of the nuclear envelope and lamina result in severe consequences, termed 'nuclear envelopathies' or 'laminopathies'. The conditions are usually inherited, are generally incurable, and include some extremely rare muscular dystrophic conditions. The rarity of genetic conditions resulting from malfunctioning building blocks of any cell component might make them appear trivial, but more likely, the searching requirements of building an organism without all the full complement of correct parts is likely to stop development proceeding much farther than a few divisions of the zygote.

The genetic constituents of the nucleus

Although the nucleus might have been recognized by Antonie van Leeuwenhoek in the late 17th century, it was not until 1831 that it was reported as a specific structure in orchid epidermal cells by a

Scottish botanist, Robert Brown (better known for recognizing 'Brownian movement' of pollen grains in water). In 1879, Walther Flemming observed that the nucleus broke down into small fragments at cell division, followed by re-formation of the fragments called chromosomes to make new nuclei in the daughter cells. It was not until 1902 that Walter Sutton and Theodor Boveri independently linked chromosomes directly to mammalian inheritance. Thomas Morgan's work with fruit flies (*Drosophila*) at the start of the 20th century showed specific characters positioned along the length of the chromosomes, followed by the realization by Oswald Avery in 1944 that the genetic material was DNA. Some nine years later, James Watson and Francis Crick showed the structure of DNA to be a double helix, for which they shared the Nobel Prize in 1962 with Maurice Wilkins, whose laboratory had provided the evidence that led to the discovery. Rosalind Franklin, whose X-ray diffraction images of DNA from the Wilkins lab had been the key to DNA structure, died of cancer aged 37 in 1958, and Nobel Prizes are not awarded posthumously. Watson and Crick published the classic double helix model in 1953. The final piece in the jigsaw of DNA structure was produced by Watson with the realization that the pairing of the nucleotide bases, adenine with thymine and guanine with cytosine, not only provided the rungs holding the twisting ladder of DNA together, but also provided a code for accurate replication and a template for protein assembly. Crick continued to study and elucidate the base pairing required for coding proteins, and this led to the fundamental 'dogma' that 'DNA makes RNA and RNA makes protein'. The discovery of DNA structure marked an enormous advance in biology, probably the most significant since Darwin's publication of *On the Origin of Species*.

We have a lot of DNA

If the double-stranded DNA in each human nucleus was laid out as a single molecule it would measure around one and a half metres in length. The genetic information it carries is stored in the order of four nucleotide bases—cytosine (C), guanine (G), adenine (A), and thymine (T)—along its length. Groups of three bases encode an amino acid (e.g. TTA encodes the amino acid leucine, TTT—phenylalanine). A single gene may require hundreds or thousands of bases to generate a single protein. In terms of the information stored along this length of DNA, it would take 200 telephone directories to print out the three billion base sequences. All the 23,600 human genes, however, fit into around 2 cm of our DNA, which leaves 98.5% unaccounted for. This was originally considered to be 'junk' DNA. The term junk is perhaps more indicative of the ignorance of early researchers, and thus 'non-coding' (i.e. not coding for genes) is a better description. As it seems unlikely that the cell should go to the trouble of replicating more than nine-tenths of its DNA each time it divides for no reason, we would do best to consider this vast majority of our DNA to have an unknown rather than no function. At least some non-coding DNA is certainly important to the cell, as damage restricted to non-coding regions has been found to be just as effective at causing cell death as that in coding regions. Non-coding DNA contains pseudogenes, sequences that are no longer used to make proteins. These may be the remains of information accumulated over an evolutionary lifetime, which may be silent for millions of years, but can be reactivated and actively transcribed. Some non-coding DNA almost certainly represents the incorporation of viral DNA from past infections. Once infected, individuals rarely completely lose virus DNA. Over evolutionary times scales, these collections could reach significant amounts, estimated at 8% for the human genome.

The genes themselves are complex structures, having a starting code (promoter) built into the beginning of each gene and an exit code (terminator) at the end. Amid the coding sequences (introns) there are intervening non-coding sequences (exons), which need to be removed before use. In general, if a primitive organism has a particular gene, then organisms of increasing complexity will contain a number of related genes in proportion to their position on the evolutionary scale. This suggests that with time, genes are often duplicated and then evolve their sequences separately.

How is DNA packaged?

To accommodate one and a half metres of double-stranded DNA within a spherical nucleus roughly one thirty-thousandth of this length, it is clear that the DNA has to be packed in a fairly sophisticated manner. Packaging such a long molecule must allow for genes to be accessible, and also for the whole of the DNA molecule to be duplicated so that exact copies can be passed to each daughter cell. At cell division, discrete blocks of DNA which exist within the nucleus but are not visible as distinct bodies, undergo further levels of coiling and supercoiling, a process called condensation. This produces the discrete chromosomes which are the familiar image of our genetic matter (Figure 8c, d, e). During final chromosome condensation, the nuclear envelope is broken down, and the chromosomes are distributed to each daughter cell (see Chapter 4 for more details). Nuclei are then rebuilt in each daughter cell, during which the rigid and rod-like chromosomes appear to lose their individual identity as they decondense and merge back into the overall structure of the daughter cell nuclei. The question of where chromosomes go in non-dividing (interphase) nuclei was answered a century after their initial discovery thanks to a technique called fluorescence *in situ* hybridization (FISH), which was developed by Joe Gall and Mary-Lou Pardue in 1969. A related technique called chromosome painting incorporates multiple fluorescent probes, allowing individual chromosomes to be recognized within the interphase nucleus. Chromosome painting shows that each chromosome occupies a distinct territory within the nucleus, usually with attachments at the nuclear lamina. Interphase chromosomes occupy about half of the internal nuclear space, the rest being filled by a host of other nuclear components, such as nucleoli and Cajal bodies (see later). The contents of the nucleus are by no means fixed, and there is a constant flux and movement of all nuclear components over both long and short distances, which requires energy.

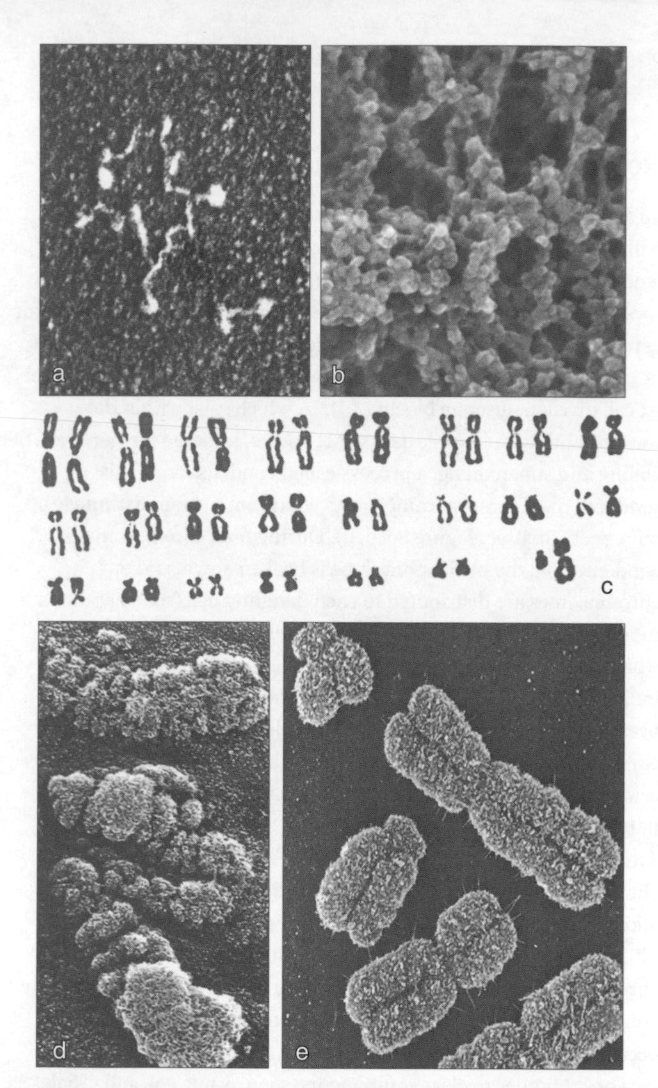

8. DNA and chromosomes. (a) Naked DNA and nucleosomes make 'beads on a string', (b) chromatin fibres within the nucleus, (c) a set of human chromosomes (known as a karyotype), (d) chromosomes during final condensation, and (e) human metaphase chromosomes

Although it is 'naked' in prokaryotes, DNA in eukaryotic cells is always associated with other molecules, and is packaged via a series of stages. Human DNA is first combined with groups of structural proteins called histones. In the first stage of packaging, DNA becomes wrapped twice around groups of eight histone molecules to form a structure known as a nucleosome, leading to a 'beads on a string appearance (Figure 8a). Adjacent nucleosomes then become attached to each other by another histone called H1, in a zig-zag fashion, forming a fibre ten nanometres in diameter. This fibre is then twisted into a solenoid configuration (a hollow tube 30 nanometres in diameter) called chromatin. Chromatin is the standard configuration of eukaryote DNA packaging (Figure 8b), and exists in two forms called heterochromatin and euchromatin. Heterochromatin is more densely packaged, producing darker staining, and tends to be peripherally distributed within the nucleus (Figure 7a). Much of the DNA in heterochromatin has short nucleotide base sequences that are repeated thousands of times (repetitive DNA) and may have a structural rather than genetic function, to anchor DNA within the nucleus. In contrast, euchromatin is much less condensed and not as densely stained, and comprises almost all of the genetically active part of the DNA. When interphase chromosomes undergo final condensation just before division, the euchromatin and heterochromatin form alternating blocks along the length of the chromosomes, which can be stained to produce a consistent banding pattern. This lengthwise series of subdivisions provides a 'road map' that has allowed individual genes to be accurately positioned not only on specific chromosomes, but to specific positions along the length of the chromosome. During final chromosome condensation, the chromatin is further looped, folded, coiled, and supercoiled, massively reducing the overall DNA length to a point where the packing ratio of DNA in a chromosome at division reaches 10,000 to 1 (Figure 8c, d, e). A good analogy to appreciate this amazing organization would be to take a skipping rope the length of a football field and fold it into an overall length of about half an inch. Recent technological advances have shown that the largest of the human chromosomes (chromosome 1) has DNA with 246 million base pairs, and disruptions in their sequence have been linked to over 350 human diseases including cancers and neurological and developmental disorders.

Now the entire genome has been sequenced, one might suspect that the chromosomes themselves might become less relevant because an individual's DNA can be analysed, compared with normal, and problems diagnosed essentially by computer. It is worth pointing out that at the time of writing only seven individuals on the planet have had their DNA sequenced. These include Craig Venter, pioneer of DNA decoding, James Watson (fittingly), two Koreans, a Chinese, a Yoruban (a member of an ethnic group from Nigeria), and a leukaemia victim. The cost of sequencing the first human genome in 2003 was around $500 million, and the most recent nearer $250,000. For sequencing to be a feasible diagnostic routine, a cost of one thousand dollars is the target, which may be technologically feasible in the near future. However, the main barrier to the medical use of genomes is that diseases such as cancer, diabetes, or Alzheimer's are invariably caused by many DNA variations, making it difficult to identify clear targets for either drug intervention or diagnostic indicators, and consequently limiting the idea of personalized medicine based on an individual genome—at least for the time being. Recently, the Wellcome Foundation have announced a project to sequence 1000 genomes, from a mixture of healthy people and those suffering from a variety of medical conditions, generating a statistically significant comparison.

The nucleolus

Separate bodies within the nucleus were first recognized by Felice Fontana in 1774 and named 'nucleoli'. Nucleoli are the largest of the distinct bodies within the nucleus and each nucleus will have up to five, clearly visible by light microscopy without need for specific staining. Nucleoli are formed from a mixture of proteins and nucleic acids, which are shown by electron microscopy (Figure 7c) to be organized into a 'tripartite' internal structure of a fibrillar centre, a dense fibrillar component, and the granular component. Nucleoli are devoted to the production of ribosomes, and the tripartite structure reflects the three events that take place there: transcription of ribosomal RNA (see Chapter 4), processing of ribosomal RNA, and ribosomal assembly. The DNA responsible for these processes is situated on five different human chromosomes at division at sites called nucleolar organizing regions (NORs). These regions come together after division to form three or four nucleoli, where the ribosomal genes are transcribed and ribosomal subunits partially assembled, ready for export from the nucleus. This concentration of genes, transcription machinery, processing, and assembly into one site allows amazing rates of production—dividing human cells make ten million ribosomes in less than a day, so the nucleolus is essentially a ribosome factory, with an efficiency that would have been the envy of Henry Ford.

Because of the demands of ribosome production, the nucleolus is also remarkably responsive to any source of stress that a cell may experience. Heat, cold, osmotic stress, and a variety of drugs all change nucleolar structure, as do viral infection and nutrient starvation. In order to see whether a cell is healthy or not, inspect the nucleolus first.

Cajal bodies, snurposomes, and spliceosomes

In 1906, Ramon y Cajal from Madrid and Camillo Golgi from Pavia shared the Nobel Prize for their work on the structure of the nervous system. Golgi discovered the apparatus or complex that took his name, while Cajal found dense staining bodies close to the nucleolus, which he originally called accessory bodies, and were subsequently named coiled bodies because of the coiled nature of their main protein, coilin. In 1999, Joe Gall suggested they should be called Cajal bodies. Cajal bodies also contain bodies called Gems and Gall's wonderfully titled spliceosomes and snurposomes (similar to coiled bodies but restricted to amphibian oocyte nuclei), all involved in the processing of RNA in the nucleus after transcription. Snurposomes contain small nuclear ribonuclear proteins (snRNPs, pronounced 'snurps'), and spliceosomes are sites where splicing of RNA takes place. In the last few years, many other intranuclear bodies have been identified, although our understanding of their biological function is still limited. Just to mention a few, there are PML bodies (also called Kremer bodies), speckles, paraspeckles, and clastosomes.

Organizing the nuclear interior

From relatively recent research, it has become apparent that far from being a mere repository for DNA, the nucleus is just as varied and dynamic as the cytoplasm in terms of content and activity. The segregation of the cytoplasm by membranes into individual organelles allows routine biochemical separation and analysis. Teasing out different components in the nucleus is trickier, as there are no such 'bordered' sub-compartments, although the high density of the nucleolus does permit its isolation from nuclear extracts fragmented by a sonic probe. From this starting point, using mass spectroscopy, some 700 human nucleolar proteins have been identified so far in a European collaboration led by Angus Lamond at Dundee University.

Interphase chromosomes occupy around half the total nuclear volume, and are separated from each other by the interchromosome space, which is filled with nucleoplasm, a viscous liquid, equivalent to the cytoplasm outside the nucleus. We now know that, as well as having their own 'domains', interphase chromosomes move around the nuclear interior. Gene-rich chromosomes (which have more euchromatin and carry the majority of active genes) tend to be found in the central parts of the nucleus, where most transcriptional activity takes place. Consequently the gene-poor chromosomes (which have more heterochromatin) are more peripherally positioned, and adjacent to the inside of the nuclear envelope where the proteins of the nuclear lamina provide a fibrous network ideal for the anchoring of nuclear contents. If the nuclear lamina is defective, then the normally anchored genetically inactive chromatin might stray into a transcriptionally active region of the nucleus and become inappropriately expressed, as happens in some of the diseases called laminopathies, such as Duchenne muscular dystrophy.

Although electron microscopy has produced vast amounts of information on the workings of the cytoplasm, it has been relatively less successful for the nucleus. This is due to the intense packaging and fibrous nature of the nuclear contents, which makes it virtually impossible to follow a length of chromatin over any distance in the thin sections required for transmission electron microscopy. Add the sheer size of the nucleus (1000 times the volume of a mitochondrion) and it would require 200–300 serial sections to be cut, collected, and photographed at around 100 images per section before any three-dimensional reconstruction could be attempted, which is currently not a feasible task. Novel approaches such as the selective removal of components can simplify things. A fibrous structure can be seen by scanning EM after the biochemical removal of DNA and chromatin. This network of fibres running through the nucleus in thicker sections is called the nuclear matrix of scaffold (Figure 7c). Although this type of approach was initially controversial because the extensive biochemical protocols during preparation might create new structures (artefacts) rather than revealing the original organization, the idea of a fibrous supporting network running throughout the nuclear interior (the nucleoskeleton) is now generally accepted.

Chapter 4
The life of cells

The cell cycle

The actual mechanics of cell division, according to Dick McIntosh at the University of Denver, require significantly more instructions than it takes to build a moon rocket or supercomputer. First of all, the cell needs to duplicate all of its molecules, that is DNA, RNA, proteins, lipids, etc. At the organelle level, several hundred mitochondria, large areas of ER, new Golgi bodies, cytoskeletal structures, and ribosomes by the million all need to be duplicated so that the daughter cells have enough resources to grow and, in turn, divide themselves. All these processes make up the 'cell cycle'. Some cells will divide on a daily basis, others live for decades without dividing. The cell cycle is divided into phases, starting with interphase, the period between cell divisions (about 23 hours), and mitosis (M phase), the actual process of separating the original into two daughter cells (about 1 hour). Interphase is further split into three distinct periods: gap 1 (G1, 4–6 hours), a synthesis phase (S, 12 hours), and gap 2 (G2, 4–6 hours). Generally, cells continue to grow throughout interphase, but DNA replication is restricted to the S phase. At the end of G1 there is a checkpoint. If nutrient and energy levels are insufficient for DNA synthesis, the cell is diverted into a phase called G0. In 2001 Tim Hunt, Paul Nurse, and Leeland Hartwell received the Nobel Prize for their work in discovering how the cell cycle is controlled.

Tim Hunt found a set of proteins called cyclins, which accumulate during specific stages of the cell cycle. Once the right level is reached, the cell is 'allowed' to progress to the next stage and the cyclins are destroyed. Cyclins then start to build up again, keeping a score of the progress at each point of the cycle, and only allowing progression to the next stage if the correct cyclin level has been reached.

Mitosis

Once the cell is ready to divide, it enters the part of the cell cycle called mitosis (M phase) which is broken down into five further phases: prophase, prometaphase, metaphase, anaphase, and telophase. The first is prophase, during which the chromosomes condense to become discrete structures. In prometaphase the nuclear membrane breaks down and the nucleoli become indistinct. The chromosomes now become further compacted, coiling and supercoiling as they condense into clearly visible paired sausage-like structures (see Figure 8c, d, e), each made of two chromatids, joined together by a structure called the centromere. The centromere provides the site of attachment to the mitotic spindle (the microtubular framework which brings about the separation of chromosome to the daughter cells) at a structure called the kinetochore. The mitotic spindle is formed from cytoplasmic microtubules, organized by a pair of centrioles which have previously replicated and migrated to either end of the cell (as described in Chapter 2). This stage completes prometaphase, which is followed by metaphase, where the mitotic spindle microtubules apply tension to the chromosomes to align them in the centre of the spindle itself, at which point they form the 'metaphase plate'.

The next stage is anaphase, in which one chromatid from each chromosome is separated and moved to opposite ends of the mitotic spindle (Figure 9). This is brought about by a shortening of the microtubules that run from the kinetochores to the spindle poles, and a lengthening of the microtubules that run from one

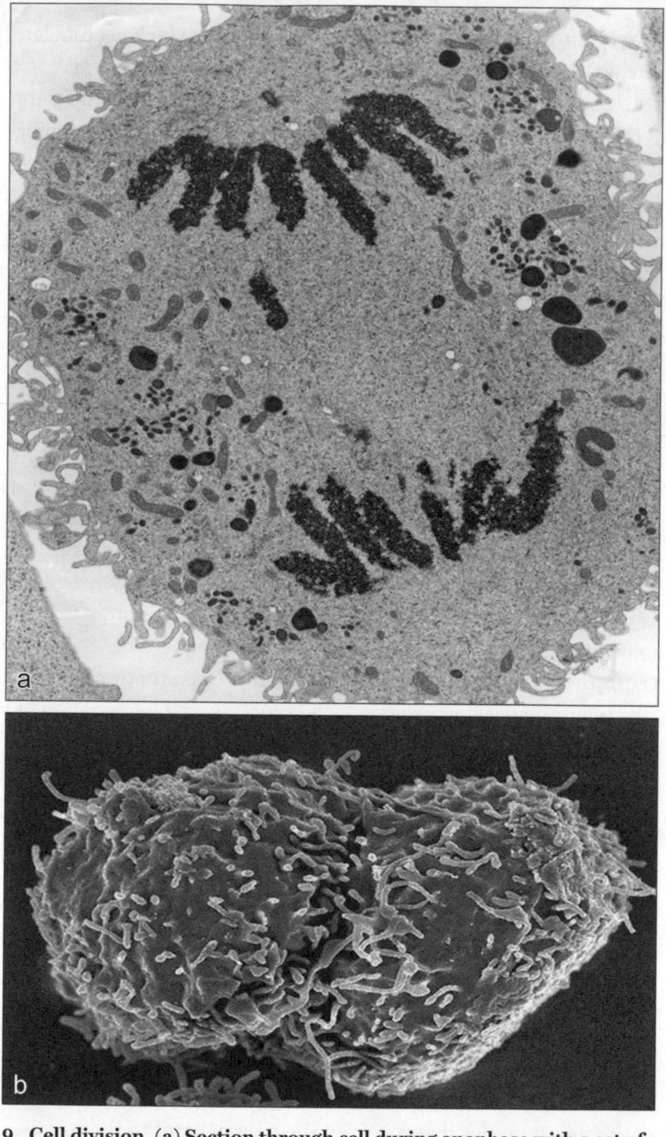

a

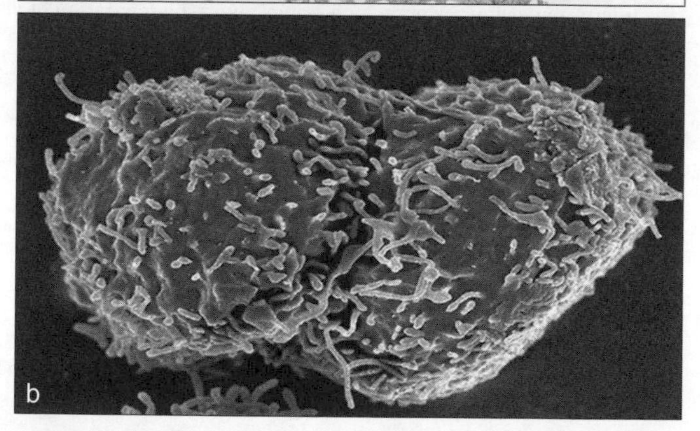

9. Cell division. (a) Section through cell during anaphase with a set of chromatids pulled towards each pole by the mitotic spindle, (b) two daughter cells beginning to separate at the end of division, 'pinching off' from each other with a central furrow end of the spindle to the other, brought about by adjacent tubules sliding over one another. The arrival of each group of chromatids at the poles of the spindle completes the last stage of mitosis, called telophase. To complete division of the whole cell into two daughter cells, a contractile ring of actin filaments forms around the middle of the cell, pinching the cytoplasm into two halves rather like a belt pulling tighter and tighter, a process called cytokinesis (Figure 9b). Once the daughter cells have separated their chromosomes will begin to relax, decondensing to their interphase configuration as part of the newly formed nucleus. A new nuclear envelope is formed at the surface of the decondensing chromosomes, which already carry the building blocks for new nuclear pores on their surfaces. Mitosis is now complete, and the cell enters interphase again, either to begin a process of further redoubling for the next division, or to leave the cell cycle (G0) and embark on a journey of differentiation to fulfil a specialized tissue function, as will be described in Chapter 5.

Meiosis

Meiosis is a reduction division, used by multicellular organisms to produce special cells (gametes) which have a single copy of DNA (haploid), in readiness for fusion with another gamete to produce a cell (zygote) with the normal two copies of DNA (diploid), from which the embryo will develop into a new organism. Gametes can be sperm and eggs as in most animals, pollen and ovules in plants, and spores in other life forms such as fungi. The creation of a new organism by fusion of gametes from two different organisms is defined as sexual reproduction. Plants, of course, are not as dependent on this process as are the majority of animals, having ways of asexual reproduction open to them. It is beyond the scope of this book to delve too deeply into the genetic consequences of sexual reproduction, but suffice to say that the constant mixing of genes brought about in this manner provides the variety which upon which the pressures of natural selection drive evolution.

The mechanics of meiosis are relatively simple: two rounds of chromosome division are completed without a round of DNA replication in between. By dividing a diploid cell twice, four haploid gametes are produced. Meiosis begins with the two matching (homologous) chromosomes (one maternal, one paternal) in a diploid cell coming together, at which point DNA may be 'swapped' in a process called crossing over. The first division separates one of each pair of chromosomes to two new daughter cells, which then divide directly producing four gametes which now have half the original DNA complement (haploid). The molecular mechanics of chromosome separation via the microtubules of the spindle in meiosis are pretty much the same as they are in mitosis. Numerically, sperm production considerably outstrips egg production, as a fertile human male produces 1000 sperm in the same time as a single heartbeat, whereas human females are born with around 500 eggs to last a lifetime.

DNA replication

Before a cell can divide, it must produce two copies of its DNA, one for each daughter cell. The two strands of DNA from the mother cell are separated, and copies are made using the original strands of DNA as a template. Remembering that the nucleotide base A always pairs with T and C with G, if one strand has a sequence ATCG then the new strand will have a sequence TAGC. The opposite old strand is TAGC and its daughter strand is ATCG. In this way, two identical copies of a DNA sequence are made. This is called semi-conservative replication, and normally provides an exact copy. Any mistakes in copying generate mutations, leading to alterations in the genetic message which are passed to the daughter cells. DNA is made in a single uninterrupted burst that takes around a third of the time that a new cell needs before it divides again (cell cycle). In bacteria, the cell cycle may last only a few minutes, and a couple of hours in simple eukaryotes such as yeast. In contrast, most mammalian cell cycles are around 24 hours. Only a small proportion of cells in the human body divide daily. For example, we manufacture a new layer of skin cells every day, and reline the surface of our intestines continuously, but some nerve cells will last a lifetime. When DNA replication is required, it takes place in around 100 'replication factories' distributed throughout the nucleus. DNA is fed in to the replication machinery like film into a projector, and emerges as two films. Exactly what happens at the molecular level could fill a book twice the size of this one, but here is a brief outline of the overall process of the transmission of genetic information between cells.

The first requirement is to separate the two strands of the DNA helix to provide templates for the formation of new strands, one on each half of the original DNA. This is (relatively) simple in prokaryotes where the DNA lies naked in the cell. In eukaryotes, it would take far too long to replicate the DNA starting at one end only, so that the DNA is opened up by an enzyme called helicase at around 1000 different sites along its length. If you consider the topology of this process using pieces of twisted string, it requires one strand of the continuous helix to be cut before a short stretch of double stranded DNA can be unwound. Then the main enzyme of replication (DNA polymerase) locks on to the opened DNA strands at these 'replication forks' and adds new nucleotide bases (in the correct order) to the new strands at a rate of 100 bases per second (Figure 10). Bacteria can do this at even greater rates—up to 1000 bases per second. Despite the rapid rate, replication is extremely accurate, with enzymes that proofread and correct any mismatched nucleotide, usually leaving only one error in every billion nucleotides.

Transcription

Transcription is the first step in the process by which genetic information generates new proteins. One strand of the DNA (known as the coding strand) is used as a template to make an RNA

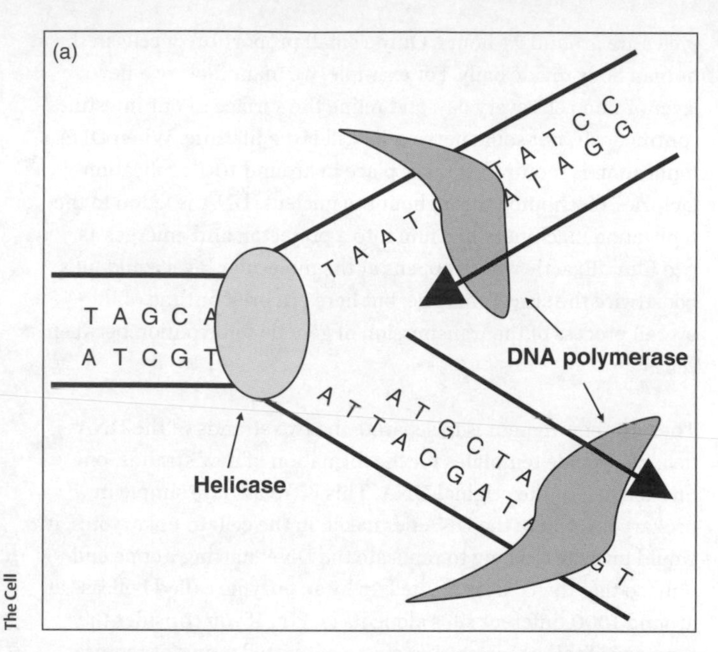

10. DNA replication. (a) A replication fork. Helicase unwinds the double stranded DNA so that each strand can be copied. DNA polymerase enzymes, one each strand, assemble the complementary strands working in opposite directions (arrows). In bacteria, the DNA strands are continuous and circular, so replication starts at one point and continues round to complete the entire DNA molecule

sequence (messenger RNA or mRNA; Figure 11). This messenger RNA will then be used as a template for protein production.

In a similar way to replication, RNA is synthesized along the DNA template but with RNA polymerase rather than DNA polymerase. Messenger RNA also uses a different nucleotide base, uracil (U), instead of the thymine (T) in DNA. Before the newly transcribed RNA passes out of the nucleus, the leading end is capped, and a tail is fixed to the trailing end. The parts of the copied DNA sequence that do not code for protein are removed by a process

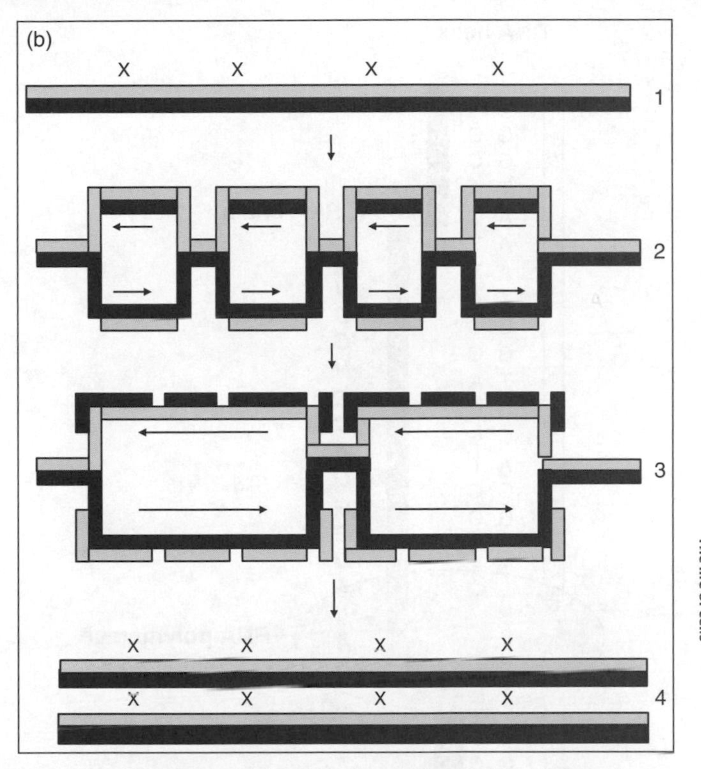

(b) In animal cells replication is started simultaneously at many points. At the top (1) is double stranded DNA (grey strand, black strand) showing the places or origins where replication starts (X). DNA polymerases begin to copy both strands in the direction indicated by the arrows (2, 3). A number of DNA polymerase enzymes replicate each strand resulting in fragments of the newly synthesized DNA strand (3). As the replication bubbles grow these fragments are linked together and checked for accuracy to make two exact copies of the starting DNA (4)

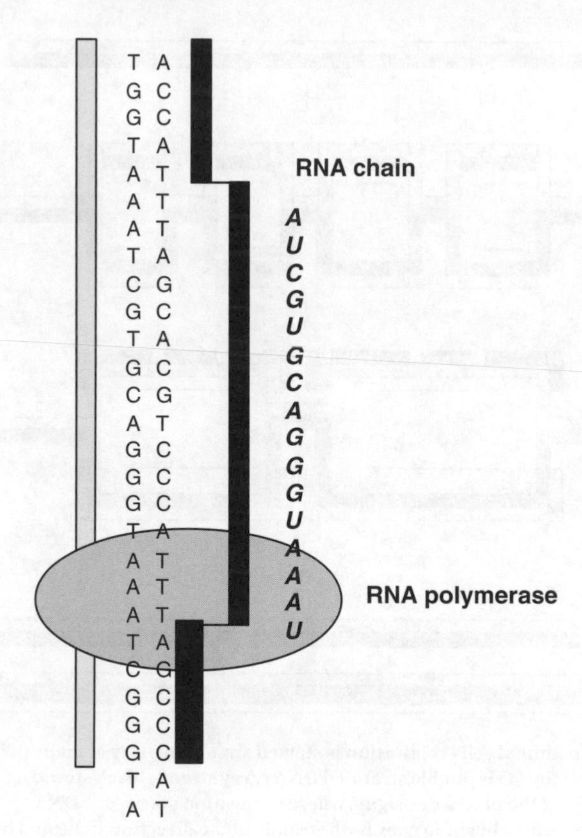

11. RNA polymerase is a large complex of proteins, which binds to specific sites at the start of genes. It unwinds the DNA helix and copies one strand to give a complementary RNA strand. The sugar molecules in RNA are different to those found in DNA. Three of the four bases (A, C, G) are the same but T is substituted for U. As the RNA polymerase moves along the DNA, its strands recombine to form a helix

called RNA splicing. The new mRNA is then tagged with proteins that will target it to a nuclear pore prior to passage into the cytoplasm, where it will combine with a ribosome to form the machinery for making a new protein.

Although the mechanics of transcription are understood, how genes are selected for transcription is less clear. The phenomenon of 'gene expression' has occupied thousands of molecular biologists for many decades. Basic genes required to maintain the cell in good working order (housekeeping genes) are always expressed but many other genes will only be required at specific times during the life of the organism. Some genes, such as the two globin proteins that make up haemoglobin, are required in such amounts that red cells switch to over 90% globin production. The cell can also switch genes on and off in a very complex pattern and the production or non-production of proteins and enzymes is the cell's main way of responding to any change of circumstance. We have around 200 different cell types in our bodies, all with specialized roles. These differentiated cells result from different genes being switched on or off in different tissues. Cells also have to be able to respond rapidly to external changes by switching genes on and off. These switches are controlled by a group of around 3000 different proteins called transcription factors. Some genes require many factors, others only a few. Transcription factors reside in the cytoplasm and must enter the nucleus to access their genes. Transcription factors required for rapid responses shuttle in and out of the nuclear pores, maintaining a constant state of readiness.

How cells move

Much of what we know of the mechanisms of how a cell moves comes from watching fibroblast cells in tissue culture. The cell moves by extending a broad leading edge, a lamellipodium (see Figure 3c) which moves by a cycle of attachment and detachment of the underside membrane to underlying surface, rather like the incoming tide moving up the beach. The upper surface of the lamellipodium has ruffles, active waves of folded membrane, which flow towards the rear of the cell. All this activity is generated under the cell membrane by the addition of subunits to the front ends of a branching network of actin filaments. At the rear of the cell, the actin filaments break down, ready to be recycled to the front of the lamellipodium. Finger-like projections termed filopodia (similar to microvilli) 'feel out' gaps between cells, allowing fibroblasts to move through solid tissue. This migration through tissues is entirely normal for both fibroblasts and white blood cells as they go about their normal duties of cell maintenance and immune defence, but at the same time represents the major problem in cancer. Tumours are initiated by a local loss of control of cell division. If the mass of newly divided cells stays put, then the resulting tumour is benign and can usually be successfully removed by surgery or killed by radiation therapy. The biggest problem with cancer is metastasis, where cells dissociate from the primary tumour, penetrate the surrounding tissue, and ultimately access the bloodstream, from which point they can generate secondary tumours virtually anywhere in the body. An understanding of exactly how cells (both normal and tumour) move through tissue barriers is the first step to find treatments which could inhibit metastasis, and new drugs which could stop the spread of cancer cell away from the original (primary) site. In the last few years, some progress has been made in identifying both genes that are active in metastasis (unsurprisingly those that are also involved in cell migration) and inhibitors of their mechanisms. Compounds from such diverse sources as citrus peel (modified citrus pectin) and olive oil (oleamides) have shown anti-metastatic properties.

Movement, an emergent property?

René Dutrochet, one of the pioneers of cell biology, remarked in 1824 that 'life, as far as physical order is concerned, is nothing more than movement; and death is simply the cessation of this movement'. Nearly 200 years later, just how cells combine the various components of the cytoskeleton to function as a single moving entity is still a bit of a mystery. Cell movement requires biochemical cues (signals), an energy supply, and the reorganization of structural elements. Cytoskeletal elements need to grow and shrink and to arrange themselves for action, but when, with how much force, and when to stop? Acccording to Guenter Albrecht-Buehler from Northwestern University in Chicago, a cell biologist studying the behaviour of cells in culture, 'the functions of the organism initiate and control the interactions between its molecules'—which is a way of saying the whole is greater than the sum of the parts. Something that happens as a result of the interaction of many complex systems is called an emergent property. In nature, the classic examples come from social insects, such as the massive cathedral-like structures produced by termite colonies, or even the production of a honeycomb. The movement of a cell could be considered an emergent property of the molecules of the cytoskeleton, supported by energy production from mitochondria, and information stored in DNA. Thus it seems feasible that the millions of cells within a tissue produce emergent properties to fulfil the purpose of the tissue, and then add another level as tissues form organs, and yet further levels as an entire organism. This view may go some way to accounting for (if not exactly explaining) the complexities of our own existence.

Is cell movement completely random?

After 30 years examining the behaviour of individual cells in culture, and recording an enormous amount of time lapse footage, the studies of Albrecht-Buehler have produced some fascinating observations which may suggest that cells both require and indeed possess 'intelligence' of their surroundings to act as they do. The idea that a cell 'knows' where it is going is reinforced by their behaviour when they meet other cells and then, after contact, move off in opposite directions. This shows a 'choice' within the cell in reaction to an unexpected event in their migration.

Albrecht-Buehler's published observations of the behaviour of cells in a culture dish appear to indicate directed movement and the apparent appreciation of other cells at a distance of several cell diameters. Albrecht-Buehler terms this 'cell intelligence', and points out that this behaviour requires incoming information, which he suggests may involve infrared light. He suggests that cells can signal to each other over a distance of several cell diameters by the emission and reception of infrared light. In the absence of any great body of work in this area of research, it is hard to assess the true significance of these studies, but they are extremely intriguing. In his own words, Albrecht-Buehler suggests that 'cell behaviour is controlled by very complex data integration systems that are, so far, unknown to biology'.

If what controls cell movement in the relative simplicity of a plastic dish is complex, then so is cell movement within the human body. Blood cells are pumped around the circulation to provide oxygenation of tissues by red cells and to maintain immune surveillance with a variety of white blood cells called leucocytes. To reach sites of injury or infection, leucocytes must leave the circulatory flow. This is achieved by attaching themselves to the wall of small blood vessels, first rolling to a stop using adhesive molecules called lectins, a bit like grappling irons, then stabilizing the attachment using stronger adhesive proteins called integrins. At this point, leucocytes migrate between the endothelial cells that line the blood vessels (rather like elbowing a way through a crowd), ready to confront invading bacteria, and maybe commit suicide by bursting to release their antibacterial contents (this will be described in more detail in Chapter 6). Once the bacteria have succumbed, their corpses are engulfed by macrophages that have also migrated to the site. The speed and specificity of this response is impressive, albeit that there are 'patrolling' cells constantly coursing through the whole blood system. There is also a regular directed migration in the production of another class of white blood cells (T cells), which leave the bone marrow as immature precursor cells and complete their development in the thymus using similar homing and adhesion molecules. Although there are general ideas about the mechanisms of migrations in the body, we are far from understanding the whole picture.

How old are our cells?

Benjamin Franklin's assertion that the only two certainties in life are death and taxes only partially applies to cells. The options for 'the end' in cellular terms are to divide or die. A cell 'avoids' death by dividing, but nevertheless sacrifices its own unique existence by becoming two daughter cells. Most daughter cells are identical, and will have identical fates, which is to differentiate to perform a particular function, eventually die, and be replaced. This may involve a lifespan of no more than a few hours, as is the case for some white blood cells (neutrophils), or up to 120 days for red blood cells. Most cells throughout our bodies will be replaced over the course of our lifetime, with an average of seven to ten years according to Jonas Frisen, from the Karolinska Institute in Stockholm. There are, however, three types of cell that we carry from the cot to the grave: the neurons of the cerebral cortex of the brain; cells in the inner part of the eye lens; and, perhaps surprisingly, heart muscle cells, which must have contracted a mind boggling three billion times in anyone who reaches 100 years of age. The cells in the inner part of the lens are structurally very similar to keratinized skin cells in that they are largely filled with keratin fibres, but laid down in a highly organized crystalline arrangement allowing the maximum amount of light to travel through them. In time, this crystalline organization may break down and the cells of the eye lens become opaque, resulting in cataracts.

Cell death

As we have seen, the life span of cells within the human body may vary between hours and decades. The actual process of death may come about in a variety of ways from a variety of causes. Direct trauma such as mechanical damage, or extreme heat or cold, produces instant and direct effects such as rupture of the cell membrane, or instant destruction of proteins. This produces a balance of disorder over order that cannot be reversed, so the cell dies, rather like the sudden and usually violent death of an individual, maybe in a road accident or war situation. In cells, this type of death is termed necrosis.

A different and more intriguing cell death is one which is a very necessary process in all multicellular organisms, originally identified in studies of development but subsequently as a 'suicide pathway' for all cells that have failed to reproduce themselves in a satisfactory manner, ending any threat posed by the continued replication of 'rogue' cells. This is known as programmed cell death, or 'apoptosis' (from Greek, 'the falling of the leaves'), first recognized in the 1970s by Alastair Currie, John Kerr, and Andrew Wylie, all pathologists in Aberdeen University, who observed (mainly by electron microscopy) a characteristic series of changes leading to cell death. It took over a decade before this process was fully accepted as a major biological mechanism for maintaining the *status quo* of cells in tissues. Apoptosis does not trigger an immune response (as can occur with necrosis), as the cell contents are effectively 'recycled', with the fragmented remnants of the dead cell taken up (phagocytosed) by their healthy neighbours. Apoptosis is routine in developmental processes such as the removal of webbing between fingers in humans, the loss of tadpole tails in amphibians, and insect metamorphosis. Cells that are no longer required during development begin to shrink, and their surfaces generate spherical protuberances called blebs (Figure 12a), a very active process when observed by time-lapse microscopy, with the surface of the cell resembling boiling mud pools in volcanic sites. Inside the cell, nuclear contents break down, and the chromatin aggregates into characteristic dense masses that became the benchmark appearance of apoptotic cells (Figure 12c), as the DNA is chopped up into small pieces.

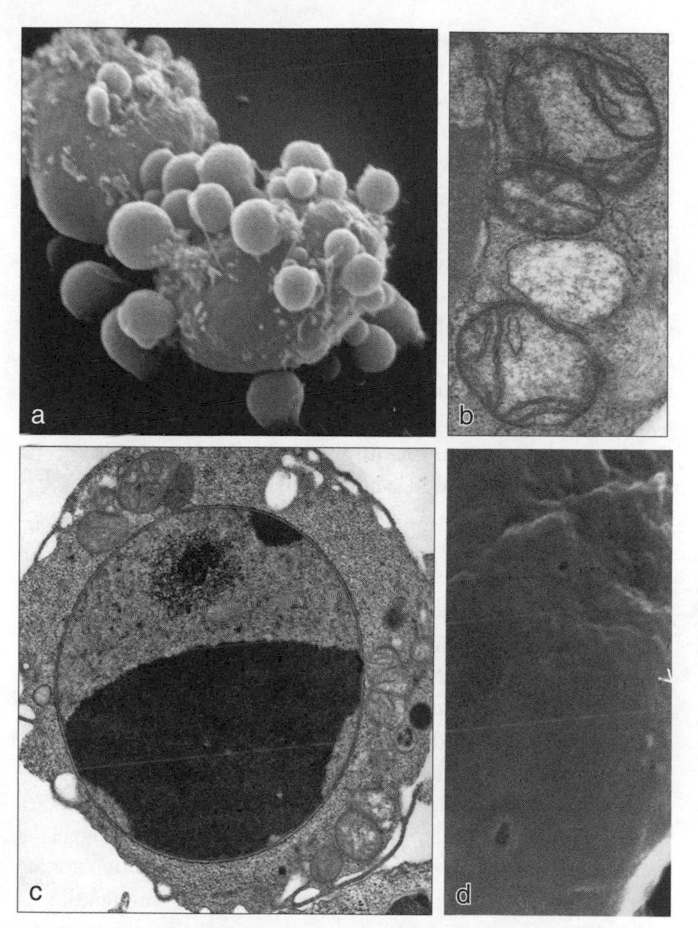

12. Apoptosis (programmed cell death). (a) Initially, the surface of a cell becomes blebbed, (b) mitochondria lose their internal structure, (c) the contents of the nucleus aggregate, and (d) pores develop in the surface of mitochondria

Once it was realized that apoptosis was a universal process in the demise of cells, the search began in earnest, to find how this 'cell suicide' was initiated. Two pathways emerged, dependent on whether apoptosis was triggered externally (extrinsic) or internally (intrinsic). Extrinsic apoptosis results from signals received at the cell membrane, which bind to 'death receptors' on the cell surface that then initiate the process of cell suicide. This type of apoptosis usually occurs as a method of cell killing in immunological reactions. Intrinsic apoptosis occurs during embryonic development where it is programmed for a particular stage or, alternatively, is triggered as a result of extensive damage to DNA, such as that caused by ionizing radiation. Apoptosis is also triggered when DNA replication is compromised and, despite the presence of repair mechanisms, the proofreading quality control systems still find inaccuracies that would lead to production of mutated proteins.

The mechanics of cell suicide

So how does a cell go about ending its own existence? The actual mechanism of 'cell suicide' depends upon mitochondria, termed the 'angels of death' by Nick Lane in his book, *Power, Sex and Suicide: Mitochondria and the Meaning of Life*. The first change occurs in the mitochondrial inner membrane, which becomes damaged by aberrant biochemical activity, leading to the formation of pores in the mitochondrial membrane (Figure 12b, d). At this point, the mitochondrion becomes committed to trigger apoptosis, and releases cytochrome *c* (a protein crucial to its normal function of energy production) which exits through the newly formed pores. This information came to light as a result of some neat experiments in which apoptotic mitochondria were introduced into perfectly healthy cells, resulting in apoptosis. The released cytochrome *c* binds to several other proteins in the cytoplasm to form a complex called the apoptosome which, in turn, activates a cascade of 'executioner enzymes' which not only kill the cell but cause fragmentation of the nucleus and cytoplasm into bite-size pieces ready to be phagocytosed by neighbouring cells.

Because intrinsic cell death occurs as part of a developmental programme, it seemed likely that there were actually genes for apoptosis. Working with *Caenorhabditis elegans* (a much studied nematode worm), where it is known that exactly 131 of the total of 1090 cells die by apoptosis during development, Bob Horovitz from Cambridge in the USA found several programmed cell death genes. These studies led to the award of a Nobel Prize in 2002, shared with John Sulston and Sydney Brenner from Cambridge in the UK.

Several of these 'cell death' genes were those genes that were routinely mutated in many mammalian cancer cells, confirming that apoptosis was a mechanism to delete cells with damaged DNA. Mutations in the cell death genes blocked apoptosis, allowing cells with damaged or mutated DNA to develop abnormally and generate tumours. David Lane, who lost his father to cancer when he was only 19, dedicated his research efforts to explore the changes that occur when normal cells become cancerous. In 1979, he found p53, a protein that is inactivated or deficient in almost all cancer cells. If normal p53 is present at the right levels in dividing cells, and a cell has damaged DNA (or replicates its DNA inaccurately), then the p53 triggers apoptosis, or, failing that, initiates a pathway to senescence, halting all division, hence inspiring its description as a tumour suppressor and the 'guardian of the genome'. Apoptotic and senescent cells are recognized by cells of the immune system (see next chapter) and removed by phagocytosis. Thus, even at advanced stages of cancer, a restoration of the 'p53 response' will bring powerful defence responses into play to shrink tumours and stop further growth. Unfortunately, getting normal p53 into cancer cells is not a simple process, and several molecular biological strategies are currently under investigation. The promise of gene therapy as the answer to multiple diseases has still to be realized, largely because the practicalities of introducing genes and their products into cells in the right amounts and at the right time without unacceptable side effects are still proving extremely difficult. Interestingly, China appears to have made the most progress in p53-based treatments, with a drug called Gendicine having been approved for head and neck cancer in 2003, whereas the US Food and Drink Administration by 2010 had not approved any p53-based treatments.

Chapter 5
What cells can do

In animals and plants, cells are grouped together in tissues and organs. Each organ is formed of a mixture of tissues which themselves contain several different cell types that work together to perform the tasks needed for the survival and reproduction of the organism. Connective tissue is characterized by large amounts of extracellular matrix secreted by well-separated cells, providing skeletal tissues such as bone, cartilage, tendons, and ligaments that make up the structural framework of the body. The more sophisticated the organism, the more complex and numerous are its cell types. In evolutionary terms, this allows for the creation of specialist cells that can respond to and survive a wide range of challenges. In this chapter we examine varied examples of cell specialisms that allow an organism to protect itself and respond to its environment.

Cells at surfaces

In both plants and animals there are common features as to how cells group and function together. Starting from the outside, there is a protective layer of cells. In plants this is called the epidermal layer, which secretes a waxy coating or cuticle that helps the plant retain water. In woody species that undergo secondary growth, the periderm, commonly called bark, replaces the epidermis and consists of cork cells giving the plant further protection from pathogens and thermal insulation. Insects are different, producing an exoskeleton made up of layers of chitin, a dense horny waterproof substance providing a protective cover which doubles up as their skeleton. Resembling a suit of armour, the exoskeleton is composed of jointed plates and the membranes connecting these give flexibility to the insect body. The organs and muscles are attached to the inner surfaces of the exoskeleton. In vertebrates (animals having a backbone), body structure is formed by an internal skeleton that is carefully laid down by a group of bone-producing cells (osteoblasts) starting early in development.

Skin in animals follows similar principles to the protective layer of cells in plants but has much more functional complexity and flexibility. All surfaces of our bodies are covered by epithelia, and we can contrast the day-to-day existence of the foot soldiers of the epithelial cells at these surfaces, on the outside and inside. Both are replaced on a daily basis, but in different ways, determined by the job that they fulfil.

Our external epithelium is called the epidermis, which when viewed from the surface is comprised of flattened plates of keratin called squames (Figure 13). Squames begin life as normal cells in the lower layers of the epidermis but, as they travel towards the surface, they progressively lose all recognizable contents, becoming plates of mainly keratin protein, based on a progressive deposition of protein on the intermediate filaments of the cytoskeleton. Because our skin is warm and moist, it provides an attractive surface that is constantly colonized by bacteria and fungi. Our response to this is to shed the outer layer of squames on a daily basis, together with the attached hitch-hikers. Shedding takes place one cell at a time, not as an entire sheet in the manner of a snake. As cells at the skin surface become detached, they are replaced by cell division in the basal layer of the epidermis. Between the basal layer and the surface (around 14 layers in humans), the differentiating cells are arranged in vertical columns. Basal epidermal cells have in their cytoplasm many bundles of keratin intermediate filaments, as well as actin and microtubule networks, which mediate changes in shape from cuboidal to flattened as the cells travel upwards, until the cell reaches the upper layers. By this time the whole of the cytoplasm has become a concentrated network of keratin filaments, interspersed with granules packed full of lipids. About halfway up, the nucleus is broken down and reabsorbed. Although the epidermal surface appears quite disorganized, if the 'loose' cells are removed, then a striking geometric arrangement is revealed, where the squames have a regular hexagonal shape (Figure 13b, c, d). The squames are not completely flat, as they have facets around their edges where they overlap with their neighbours. Two large flat surfaces and twelve edge facets make each squame a fourteen-sided solid or tetrakaidecahedron. This is exactly the same minimum area surface configuration adopted by bubbles in a stable foam. This demonstrates that the shape of cells follows the laws of physics, ensuring the maximum surface coverage for the minimum use of material for each squame. This arrangement also guarantees that the cells at the top fall off as individual cells, because each cell is only free to detach when all its six neighbours have gone, and its own six edges are free (Figure 13d). Loss of surface cells one at a time maintains a constant overall thickness, with no tearing which might allow bacteria to penetrate more deeply. Sharp edges and solid geometry are not the first thing one associates with cells, but are there as the most efficient solution for maintaining our external surfaces and as a result of the meeting of natural selection with the laws of physics.

While skin provides a remarkably efficient watertight and mechanical barrier to the external environment, these parameters are exactly the opposite of the requirements of gut epithelium, where we need to optimize our uptake of nutrients, while at the same time inhibiting the uptake of anything potentially harmful. Anything we swallow is subject to a journey of around 30 metres over the course of 35 hours. Ingested food is subjected first to the

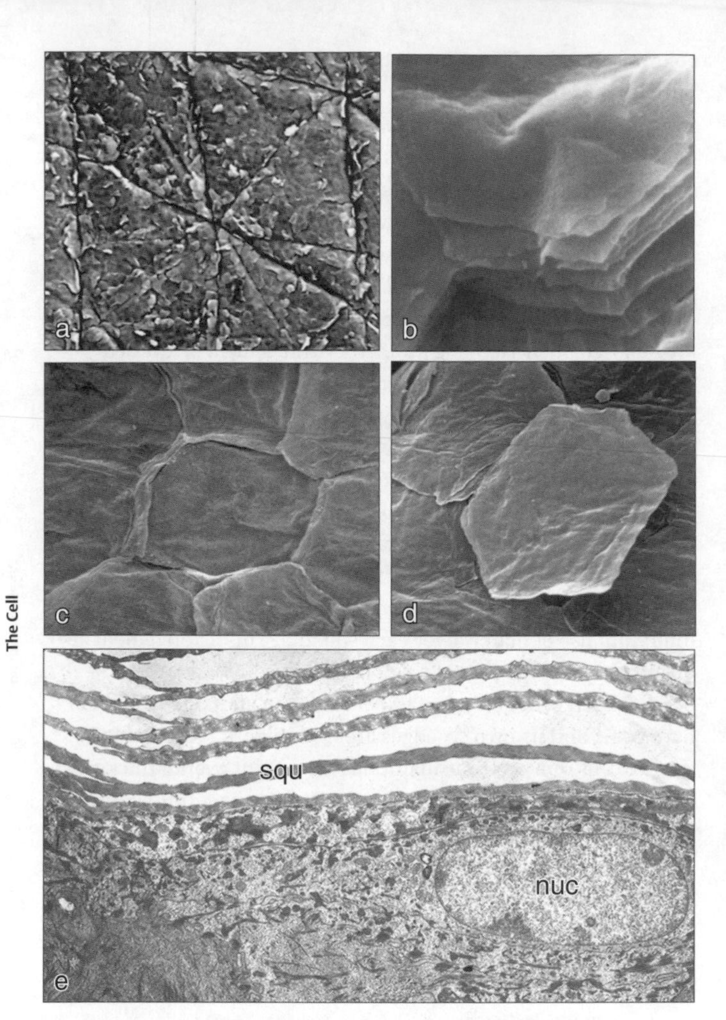

13. Surface views of skin cells at increasing magnifications (a–c), showing the stacked, hexagonal organization, and release of a single squame (d), (e) section through the stack of cells, with the last stage of a cell with a nucleus (nuc) before differentiation into squames (squ) above

ood cells

ood is classified as a connective tissue (like bone), but with
uid rather than solid matrix. This fluid transports nutrient
moves waste around the animal. Insects and crustaceans of
ave yellow or green 'blood' as they absorb oxygen directly int
heir tissues through small tubes throughout their body and t
lood lacks our own red oxygen-carrying protein called
haemoglobin.

Blood is formed from a collection of various cell types, which
continuously produced from a small number of stem cells (cel
capable of developing into many cell types). Our understandin
how blood cells develop has acted as a model for many other t
systems. In adult mammals, blood stem cells reside in the bon
marrow (Figure 15) and are characterized by a 'self renewing'
division, in which half of their daughter cells remain as stem c
with the other half becoming progenitor cells. These progenito
cells divide many thousands of times, progressively undergoin
process of differentiation that changes biochemistry, shape, an
size until, finally, a mature blood cell is produced. The site of
human blood cell formation changes throughout life. In embry
it occurs in aggregates of blood cells in the yolk sac (called bloc
islands). As development continues, blood is produced in the
spleen, liver, and lymph nodes. When the bones develop, blood
production begins in the marrow of juvenile long bones (thigh
shin) but in adults it moves to the marrow of the pelvis and
sternum. Humans make around 150 billion new blood cells per
hour. Most are the red cells that carry oxygen around the body.
Immature red cells develop into their final stage (erythrocytes)
under the influence of a protein growth factor (erythropoietin c
Epo) switching their protein production over to the two globin
genes needed to form haemoglobin (Figure 15b, c). Red cells liv
for a little over 100 days in the body. If you live or train for spor
at high altitude where oxygen levels are lower, the production o
Epo is stimulated and your blood contains more red cells. One

strongly acidic environment in the stomach, followed by powerful enzymes in the small intestine, then fluid absorption in the large intestine, and the final evacuation of waste material. We will concentrate on the small intestine, where most nutrient uptake occurs. Here, in total contrast to the multiple layers of cells making the barrier of skin, the cells form a layer that is just one cell thick, directly above the network of tiny blood vessels (capillaries) that absorb nutrient material directly into the bloodstream. Patrolling this single cell boundary and guarding against potential entry into the bloodstream by gut dwelling bacteria, is the gut-associated lymphoid tissue. This part of the immune system produces more immune cells (see later in this chapter) than anywhere else in the body at specialized areas called Peyers patches, reacting to any potential threat in the gut contents (Figure 14a). This is a crucial part of our digestive system, because around 1000 different species of bacteria reside in our gut, which together outnumber the total number of cells in our body by a 1000 fold. Most gut bacteria are harmless or even beneficial, and are tolerated by the powerful immune system in the intestine. When we inadvertently ingest harmful organisms, these are dealt with by the large numbers of mononuclear cells stimulated by antigens already in the gut and keeping the gut in a state of defensive readiness, which is called 'physiological inflammation'. In most cases, after a few days of unfortunate symptoms, things return to normal. Thus, there is a delicate balance between tolerance and immunity. Over-reaction by our immunological defences leads to allergies and food intolerances, or more serious conditions such as irritable bowel or coeliac disease.

The vast majority of the cells lining the small intestine are called enterocytes, although there are other cell types with important functions. Goblet cells secrete the mucus that covers the surface of the entire small intestine. Paneth cells reside in the epithelial crypts (of which more later) and secrete a variety of anti-microbial enzymes. Without the uptake function of the small intestine we

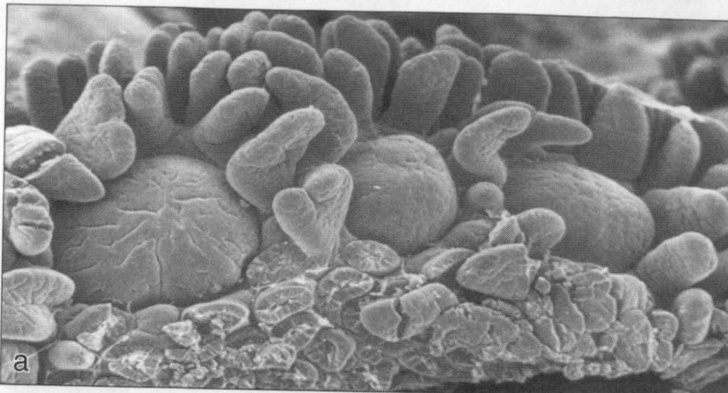

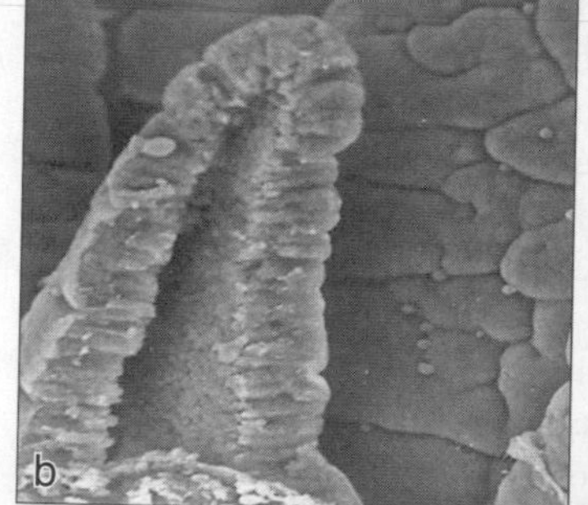

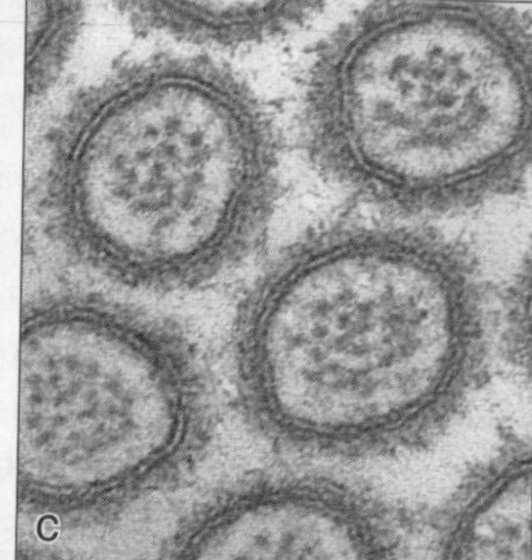

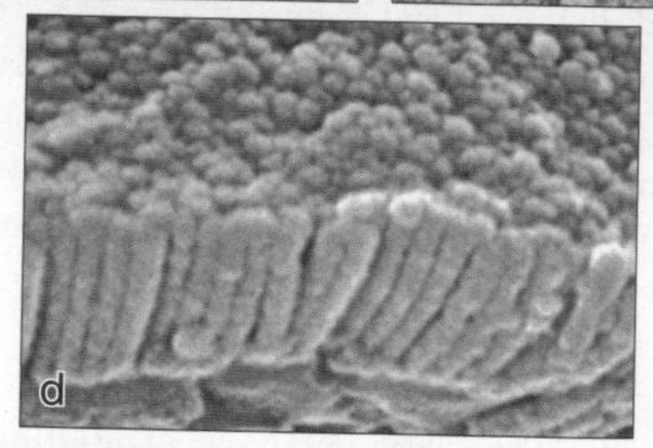

14. Gut epithelium. (a) Surface view of villi surrounding three dome-like Peyers patches, (b) a fractured villus, showing its single cell thickness, (c) section through a microvillus, showing the two layers of the membrane and internal actin filaments, (d) surface and edge view of microvilli

would very quickly starve, and without the anti-mi
we would rapidly die of infection.

In order to maximize the surface area available for ab
gut lining is neatly organized into finger-like projectio
villi, each made of around 2000 cells (Figure 14). Each
has its luminal surface organized into microvilli (Figure
supported by a central core of actin filaments, forming a '
border'. These specializations increase the area of uptake i
small intestine to something like an entire football field. Gu
contents are absorbed across the brush border of the entero
and then diffuse across the capillary walls and into the
bloodstream. All this frenzied metabolic activity, as well as
exposure to the continual threat of bacterial invasion, has resul
in the life of an enterocyte being a short one, with each cell lasti
no more than two or three days before replacement. The supply o
new enterocytes comes from invaginations rather like small
pockets found at the base of the villi, called crypts. Each villus has
five to ten crypts where precursor cells divide at a rate that
produces around 1400 cells each day. These new enterocytes
migrate towards the tip of the villus, where, having usefully
absorbed nutrient for a couple of days, they are then shed at a rate
of one every minute. This continual (lifetime) replacement of cells
adds up to a staggering production that is equivalent on a yearly
basis to about three times the body weight (as worked out in
mice). The exact mechanisms of this mass migration are still
unknown, although there are several theories, including the idea
of pressure generated by cell division in the crypts literally forcing
cells upwards. Alternatively, the cells may migrate up the
basement membrane in much the same way as migration
elsewhere. Having spent their short (but useful) life at the top of
the villus, the enterocytes lose attachment to the basement
membrane, at which point the detached cells get ejected by
crowding of surrounding cells rather like grasping and squeezing
a bar of soap until it shoots out of your hand.

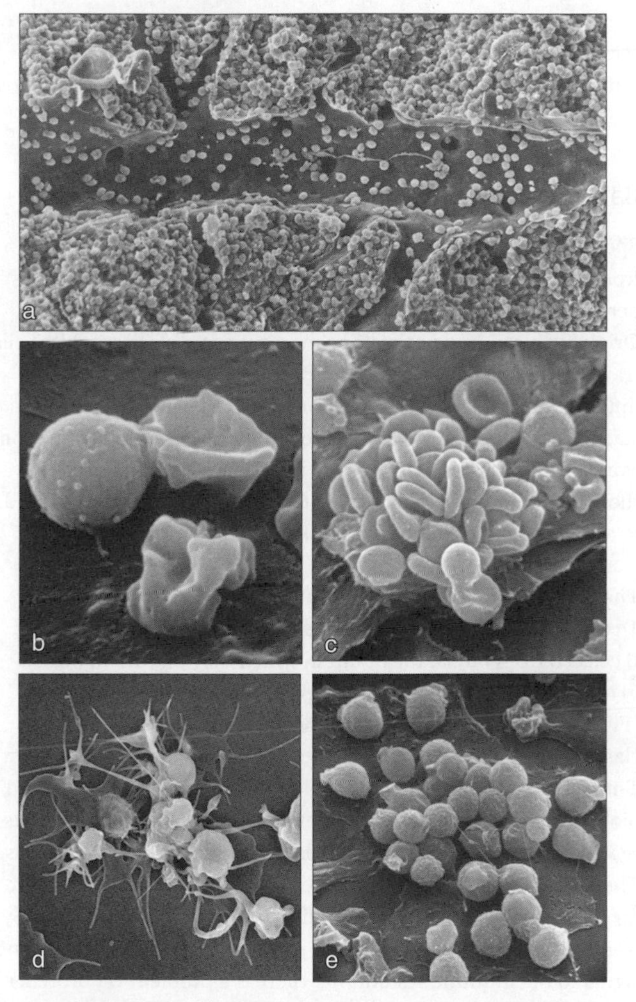

15. Blood cells. (a) Cells newly formed in the bone marrow, entering the bloodstream at the central venous sinus, (b) a developing red cell losing its spherical nucleus, (c) a group of developing red cells on the surface of a macrophage, (d) platelets make long extensions to form a clot, and (e) white cells in cultured bone marrow

the biggest challenges to keeping sport 'clean' has been the scientists' ability to discover whether sportsmen have gained unfair advantage from the artificial administration of Epo to boost their performance.

Battling and defending cells

Specialist cells have evolved to protect multicellular organisms from attack by bacteria, viruses, and parasites. White blood cells perform these defensive and cleaning-up roles within the body. Originally produced in the bone marrow, some of these white cells travel to the spleen, thymus, and lymph glands for their final differentiation. Response times for white cell production are staggering. For example, if we catch flu, our white cell production can be tripled within hours of the infection. Understanding how these blood immune cells protect the body from infection has led to a vibrant area of biomedical sciences called immunology.

The cornerstone cells of the immune system are lymphocytes or T cells (so named because these cells mature in the thymus) and B cells (B from bursa, the organ in which they mature in birds; in other animals the B cell develops in the bone marrow) (Figure 15a). T cells are further divided into various classes—memory, helper, cytotoxic, and regulatory or suppressor. T-memory cells carry information about harmful substances and biological infections (known as antigens), long after the body has resolved the infection. They can also be triggered by inoculations. These cells often survive for a lifetime and continuously monitor for the presence of antigens specific to types of infection. Upon re-exposure to this antigen, the T-memory cells rapidly divide and signal to the immune system that there is a problem via proteins on their cell surface. Given the massive numbers of potential antigenic sequences, which can be as small as two or three consecutive amino acids in a protein, a chemically modified sugar molecule, or even the shape of the protein, there are millions of different T-memory cells.

Regulatory T cells are important in the process known as immunological tolerance, which damps down or suppresses reactive T cells if they mistake your own proteins as foreign following an immune reaction. Helper T cells stimulate the growth of other immune cells. If the other T-cell types are the diplomats and generals of the immune system, then cytotoxic and natural killer T cells are the infantry. They recognize virally infected and tumour cell targets and kill them by injecting the target cell with proteins that induce cell death.

Mature B cells produce complex proteins called immunoglobins which combine to form antibodies, each capable of binding to a specific molecular structure. The necessarily massive repertoire of antibodies is created by splicing separate pieces of the large immunoglobin genes together to make immunoglobulin proteins with different active sites, each capable of binding to just one specific antigen (one small part of a protein or sugar structure).

The other major group of white cells are the myeloid cells, more varied in structure than T and B cells and comprised of granulocytes, megakaryocytes, and macrophages. All are involved in immunity and the removal of foreign biological material by different mechanisms to that of the T and B cells. Granulocytes, which contain granules, are further subdivided into neutrophils, eosinophils, and basophils. One litre of human blood contains about five billion neutrophils (around half of all white blood cells). If neutrophils receive a signal from a site of injury, it takes around 30 minutes for them to leave the blood and reach the site of potential infection. Neutrophils are serious killing cells, ferociously engulfing invading bacteria that have been targeted by an antibody and then digesting their target. Once their job is finished they turn into pus cells. The less common eosinophils (from the Latin for their 'acid loving' reaction to chemical dyes) are responsible for destroying parasites by injecting them with hydrogen peroxide (commonly used as a hair bleach and disinfectant). Fortunately, eosinophils in the circulation can only live for a few hours when activated. Eosinophils are mediators of allergic responses and are active in the development of asthma. They are also involved in many other biological processes, including graft rejection and cancer. Basophils (they react with alkaline chemical dyes) are found in great numbers at the site of parasite infection, for example ticks. Mast cells were first observed by Paul Ehrlich in the late 19th century. The large granules they contained led to the idea that they existed to feed the surrounding cells, and thus Erlich named them *Mastzellen* (from the German *Mast,* meaning food). Their granules contain histamine which, when released, triggers allergic reactions such as hay fever.

Megakaryocytes are ten times larger than red blood cells and each produces large numbers of the small platelets that are responsible for blood clotting. They normally account for 0.001% of bone marrow cells. As they mature, the DNA is replicated several times but the cell does not divide, a condition known as polyploidy which allows cells to increase in size. Some cells have as many as 64 copies of their DNA (a normal cell contains just two copies). At this stage, the cell matures and, in response to the protein hormone thrombopoietin, produces 'pro-platelet' bodies. The whole megakaryocyte undergoes a controlled 'explosion', fragmenting into a few thousand platelets, which often persist as ribbons in the blood vessels. Humans produce a billion platelets per day having a life span of four to five hours. Platelets (the only cells besides red blood cells that lack a nucleus) aggregate as clumps at the endothelial cell lining to prevent blood loss from damaged blood vessels. As they are activated to aggregate they change shape, extending long finger-like processes that interlock with each other (Figure 15d). The remaining myeloid cells are large lumbering amoeba-like cells called macrophages. Their role is rubbish collection. They engulf cellular debris and pathogens and swallow them whole (a process known as phagocytosis) before breaking their components down for further use. The average macrophage can digest a hundred bacteria before bursting as it succumbs to its own excesses. Foreign material is phagocytosed strongly acidic environment in the stomach, followed by powerful enzymes in the small intestine, then fluid absorption in the large intestine, and the final evacuation of waste material. We will concentrate on the small intestine, where most nutrient uptake occurs. Here, in total contrast to the multiple layers of cells making the barrier of skin, the cells form a layer that is just one cell thick, directly above the network of tiny blood vessels (capillaries) that absorb nutrient material directly into the bloodstream. Patrolling this single cell boundary and guarding against potential entry into the bloodstream by gut dwelling bacteria, is the gut-associated lymphoid tissue. This part of the immune system produces more immune cells (see later in this chapter) than anywhere else in the body at specialized areas called Peyers patches, reacting to any potential threat in the gut contents (Figure 14a). This is a crucial part of our digestive system, because around 1000 different species of bacteria reside in our gut, which together outnumber the total number of cells in our body by a 1000 fold. Most gut bacteria are harmless or even beneficial, and are tolerated by the powerful immune system in the intestine. When we inadvertently ingest harmful organisms, these are dealt with by the large numbers of mononuclear cells stimulated by antigens already in the gut and keeping the gut in a state of defensive readiness, which is called 'physiological inflammation'. In most cases, after a few days of unfortunate symptoms, things return to normal. Thus, there is a delicate balance between tolerance and immunity. Over-reaction by our immunological defences leads to allergies and food intolerances, or more serious conditions such as irritable bowel or coeliac disease.

The vast majority of the cells lining the small intestine are called enterocytes, although there are other cell types with important functions. Goblet cells secrete the mucus that covers the surface of the entire small intestine. Paneth cells reside in the epithelial crypts (of which more later) and secrete a variety of anti-microbial enzymes. Without the uptake function of the small intestine we

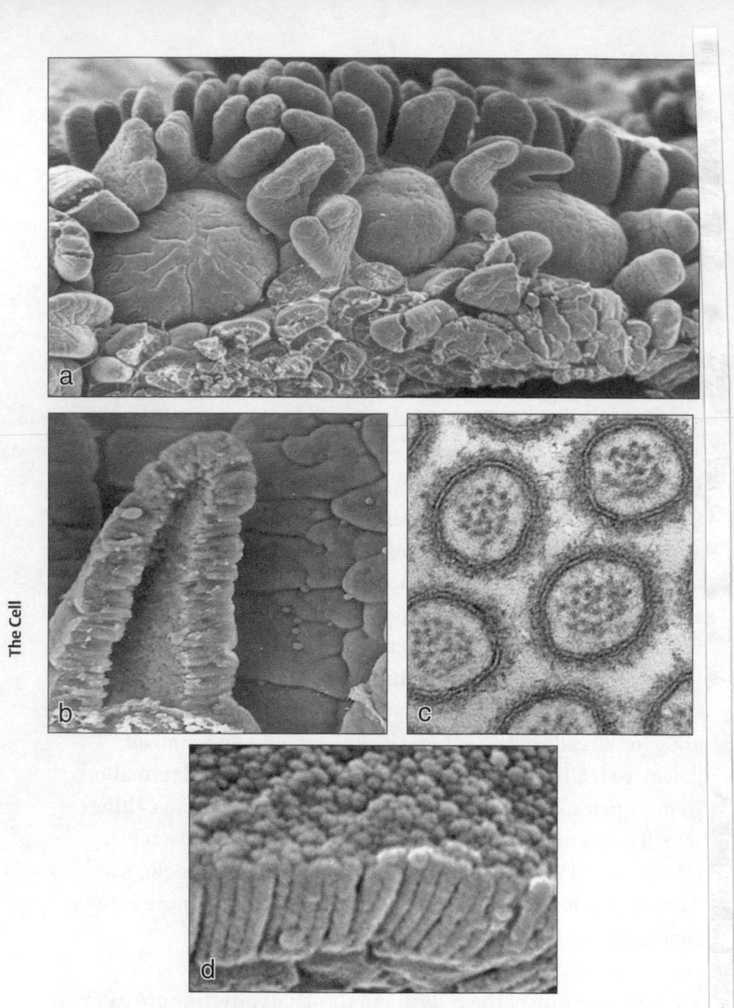

14. Gut epithelium. (a) Surface view of villi surrounding three dome-like Peyers patches, (b) a fractured villus, showing its single cell thickness, (c) section through a microvillus, showing the two layers of the membrane and internal actin filaments, (d) surface and edge view of microvilli

would very quickly starve, and without the anti-microbial barrier, we would rapidly die of infection.

In order to maximize the surface area available for absorption, the gut lining is neatly organized into finger-like projections called villi, each made of around 2000 cells (Figure 14). Each villus cell has its luminal surface organized into microvilli (Figure 14c, d) supported by a central core of actin filaments, forming a 'brush border'. These specializations increase the area of uptake in the small intestine to something like an entire football field. Gut contents are absorbed across the brush border of the enterocytes, and then diffuse across the capillary walls and into the bloodstream. All this frenzied metabolic activity, as well as exposure to the continual threat of bacterial invasion, has resulted in the life of an enterocyte being a short one, with each cell lasting no more than two or three days before replacement. The supply of new enterocytes comes from invaginations rather like small pockets found at the base of the villi, called crypts. Each villus has five to ten crypts where precursor cells divide at a rate that produces around 1400 cells each day. These new enterocytes migrate towards the tip of the villus, where, having usefully absorbed nutrient for a couple of days, they are then shed at a rate of one every minute. This continual (lifetime) replacement of cells adds up to a staggering production that is equivalent on a yearly basis to about three times the body weight (as worked out in mice). The exact mechanisms of this mass migration are still unknown, although there are several theories, including the idea of pressure generated by cell division in the crypts literally forcing cells upwards. Alternatively, the cells may migrate up the basement membrane in much the same way as migration elsewhere. Having spent their short (but useful) life at the top of the villus, the enterocytes lose attachment to the basement membrane, at which point the detached cells get ejected by crowding of surrounding cells rather like grasping and squeezing a bar of soap until it shoots out of your hand.

Blood cells

Blood is classified as a connective tissue (like bone), but with a liquid rather than solid matrix. This fluid transports nutrients and removes waste around the animal. Insects and crustaceans often have yellow or green 'blood' as they absorb oxygen directly into their tissues through small tubes throughout their body and their blood lacks our own red oxygen-carrying protein called haemoglobin.

Blood is formed from a collection of various cell types, which are continuously produced from a small number of stem cells (cells capable of developing into many cell types). Our understanding of how blood cells develop has acted as a model for many other tissue systems. In adult mammals, blood stem cells reside in the bone marrow (Figure 15) and are characterized by a 'self renewing' division, in which half of their daughter cells remain as stem cells, with the other half becoming progenitor cells. These progenitor cells divide many thousands of times, progressively undergoing a process of differentiation that changes biochemistry, shape, and size until, finally, a mature blood cell is produced. The site of human blood cell formation changes throughout life. In embryos, it occurs in aggregates of blood cells in the yolk sac (called blood islands). As development continues, blood is produced in the spleen, liver, and lymph nodes. When the bones develop, blood production begins in the marrow of juvenile long bones (thigh and shin) but in adults it moves to the marrow of the pelvis and sternum. Humans make around 150 billion new blood cells per hour. Most are the red cells that carry oxygen around the body. Immature red cells develop into their final stage (erythrocytes) under the influence of a protein growth factor (erythropoietin or Epo) switching their protein production over to the two globin genes needed to form haemoglobin (Figure 15b, c). Red cells live for a little over 100 days in the body. If you live or train for sports at high altitude where oxygen levels are lower, the production of Epo is stimulated and your blood contains more red cells. One of whole and its membranes are biochemically degraded, producing small 'foreign' protein fragments (antigens). These are then exported to the outer surface of the macrophage where they are 'presented' to T cells. Once the T cell has memorized this protein sequence, it rapidly divides, stimulated by a growth factor produced by the macrophage.

What has become clear by examining the types, the functions and molecular mechanisms of immune cells, especially in mammals, is that all are complex, interlinked, and sometimes redundant in function. This allows the body to cleanse itself of infection and to cope with most errant cells that present a threat to the organism.

Responding to the physical world

How have cells evolved to react and respond to external physical, chemical, and biological effects? Not surprisingly, cells both respond to, and make use of the environmental factors of light and gravity. Photosynthesis was established some billion years ago by ancient bacterial precursors of modern cyanobacteria. These photosynthesizing organisms were engulfed by early cells to produce chloroplasts, leading to the evolution of plants. Photosynthesis is a two-stage process involving the chemical trapping of light energy and its conversion to complex carbon–carbon bonded substances such as sugars. These sugar molecules can be used by the plant for growth but also, either directly or indirectly, act as a food source for the growth of all the non-photosynthesizing organisms on the planet. Bacteria of the deep sea and also deep terrestrial cave systems are the only cells that survive in the total absence of sunlight, as they use volcanic or inner earth heat as their energy source and have a biochemistry based on sulphur rather than oxygen. As we discover more about the processes that occurred in the first billion years of Earth's history, it seems that all bacteria must have sustained themselves in the absence of oxygen. As things cooled down between three and four billion years ago, aquatic bacteria were able to use carbon dioxide in the sunlit oceans to produce oxygen and complex carbohydrates. The rise in atmospheric oxygen content increased in several stages, beginning with a significant rise about 2.5 billion years ago known as the 'Great Oxygenation Event', although it was only after further rises following the melting of the so-called 'Snowball Earth' glaciers that oxygen levels were sufficient to sustain multicelluar animals. The chemical conditions for the majority of complex life forms were established by about 600 million years ago, perhaps the most truly earth-changing event.

All living organisms respond to light (phototropism) and gravity (gravitropism). The upper parts of plants usually grow away from gravity and towards light. The effect of earth's gravitational pull on plants was elegantly shown when mosses grown on the International Space Station grew in a spiral fashion instead of their normal upright structure. The growth control mechanism in higher plants relies on the presence of tiny dense starch-filled particles (amyloplasts) found free within the cytoplasm of specialized cells called statocytes. These cells are found in the column of the root and the joining layer of the shoot. Gravity normally pulls these particles downwards in the cell but in microgravity conditions of space they 'float', and fail to produce the usual growth pattern. The normal sedimentation of amyloplasts, within a cytoskeletal network of fine actin filaments, is thought to activate a molecular signalling pathway that leads to the redistribution of the plant hormone auxin. Changing the levels of auxin within cells is complex, as it can promote cell elongation in the shoot but inhibit it in the root.

Cells working together

We can learn much about how cells work with each other from studying simple animals such as *Caenorhabditis elegans*, a roundworm the size of a comma on this page, less than one millimetre long. The developmental biologist Lewis Wolpert may have called it 'the most boring animal imaginable' but the phylum of nematoda to which *C. elegans* belongs probably makes up 80% (numerically) of all animals in the world. Although it possesses no brain it does have a simple nervous system, a wiggling mechanism, a digestive tract, and egg-laying capacity. As mentioned in the previous chapter, it begins life with 1090 cells, but 131 are eliminated by apoptosis during development. These nematodes enjoy a soil-borne life of eating bacteria and reproducing usually with themselves as hermaphrodites by sequentially producing sperm and then eggs (only 1 in 2000 worms are truly male). For its size, *C. elegans* has a surprisingly large number of genes, around 20,000 (humans have around 24,000). Many of its genes are involved in cell division and 50% of them are shared with the banana. Over a third of its genes have a direct human counterpart. How do we explain this high number of genes in such a simple organism? One suggestion is that there are an inflated number of chemical receptor genes, allowing the worm to efficiently detect many different smells when hunting its bacterial food. Nematode worms are found in every imaginable soil type and climate. To survive in these widely different ecological niches, nematodes have evolved by collecting more and more protective or adaptive genes to allow them to survive the challenges made by all the varied species of soil bacteria, fungi, and other microbes on which they dine. The development of *C. elegans* is characterized better than any other multicellular organism, thanks to the Nobel Prize winning work of Sydney Brenner and his team, and we now understand precisely which cell develops into every other cell, including how all the 302 nerve cells connect to each other.

Moving up through the evolutionary scale, the cell development of the fruit fly, *Drosophila melanogaster*, has also been carefully mapped. Within three hours of fertilization, the fly embryo cells already start to show the first visible signs of differentiation, and the particular organ or tissue that develops depends on the exact position of each cell. Just how various organs and tissues, including brain, blood, fat tissue, thorax, and retina, develop from the embryo has been carefully established. With its extremely well defined genetics, this model organism has helped us to understand exactly how the entire cell lineage of the insect is laid down. In the next chapter, we examine how these embryo-derived cells give rise to all the adult cells found in the body of an organism.

The nervous system

Most organisms have a mechanism of movement to find food or escape from danger. The simple flagellum on a bacterium allows it to swim, whereas insects, fish, and mammals have sophisticated and coordinated sets of muscles, tendons, and nerves that can move the entire body with amazing speed and agility. Feeling and reacting to external stimuli begins with the network of nerve cells. The nematode worm has no brain but does have an interconnected collection of responsive nerve cells that performs a basically similar function to the massively complex organization of the human brain and nervous system.

The human brain contains around ten billion neurons. Each cell forms connections with thousands of others, allowing the brain to store and transmit information by passing electrical signals within the cell and chemical signals (neurotransmitters) between cells in a complex network extending throughout the body. Millions of sensory neurons have receptors that convert stimuli from the environment (light, touch, sound, smell) into electrical signals that feed back to the brain. Other motor neurons send information from the brain to the muscles and hormone-secreting glands. Interneurons mediate the information between the sensory and motor neurons. Neurons have specialized extensions called dendrites and axons. Dendrites bring information to the cell body and axons take information away from the cell body. Extended axons run large distances and are surrounded by a specialized structure called a myelin sheath, made up of cells that wrap the axon in multiple layers of membrane. The myelin sheath acts as insulation to promote the passage of nerve impulses and isolate the nerve axon from external interference. Synapses are the junctions between the cells where chemical or electrical signals are transferred. Neurons are our longest and most long-lived cells. Corticospinal neurons (which connect the motor cortex to the spinal cord) and primary afferent neurons (which extend from the organs such as skin and gut into the spinal cord and up to the brain stem) can be several feet long. Neurons last a lifetime but numbers do decrease with age. Glia cells are closely associated with all neurons, having a protective and nurturing role. We cannot understand the complex functions of vision, consciousness, and memory at the single cell level but only in terms of interconnections between many billions of cells, and perhaps the most extreme emergent property in biology. Important insights have been provided by the continuing study of the nervous system in nematodes, fruit flies, and other simple organisms which will, in time, shed light on how our own brain functions.

Cellular change

Over time, all cells accumulate multiple genetic changes to their DNA sequence. These changes are often the result of radiation damage (ultraviolet light, cosmic rays, radioactivity) or exposure to low levels of toxic chemicals. As they occur at random, the majority will miss the 2% of coding (genetically active) DNA, and have no major effect on the organism. Other rarer mutations that result in useful adaptations may arise by the alteration of single amino acids of a particular protein. This can modify the three-dimensional structure of the protein which can raise, lower, or negate its normal activity. Other changes can result in either complete or partial deletion of genes (including gene control sequences) or a subtraction or duplication of genes resulting in potentially lower or higher amounts of a particular protein. In rare cases, a major rearrangement of the DNA in the nucleus may result in a completely novel hybrid gene. This is the mechanism of evolution at its simplest and if such genetic changes adversely affect a single-cellular organism it dies. Should an advantage be conferred, it will spread throughout a population and eventually become established.

Multicellular organisms are subject to the same evolutionary pressure through DNA changes but each tissue or organ has billions of cells and at any moment of time only a proportion will be subject to genetic insult or change resulting in a disease. For example, if such a change results in a higher rate of cell division then, in time, these cells will eventually overgrow their normal counterparts (as seen in tumour growth). Adaptation relies on more subtle changes in particular cells to respond to environmental changes. Within the complex genome of most organisms there are alternative multiple pathways of proteins which can help the individual cell survive a variety of insults, for example radiation, toxic chemicals, heat, excessive or reduced oxygen. Often this is performed by repairing the damage or slowing the activity of the cell-growth machinery, waiting for the insult to go away. Many of these responsive biochemical pathways have evolved from those found in single cell genomes whose genetic make-up comes from a distant past when the Earth's climate was far more extreme. In some circumstances, the cells cannot repair damage and die through apoptosis (as outlined in the last chapter). With time, disabling mutations accumulate and coping with their consequences proves too much for the organism. If the result of these faults is tissue or organ failure, this accelerates the demise of the organism. Only if the majority of a population fails to adapt does a species die out. Changes in 'the cell' as a driving force in evolution can really only be conjecture—dinosaurs were built of the same building blocks as ourselves, yet they disappeared dramatically, having spent infinitely longer dominating the earth than our own miserable 10,000 years. It would seem that the increasingly unsteady edifice of emergent properties from combining cells into tissues, tissues into organs, and organs into creatures itself makes the end product infinitely more sensitive to sudden environmental change than its individual basic building blocks. After all, we know that single cells (sperm and eggs) can be successfully frozen for decades, whereas those individuals who had their corpses (or just heads) frozen in the hope of future revival are merely wasting their money.

Chapter 6
Stem cells

How are the billions of cells in our bodies made? In plants and animals, there are cells capable of producing every type of cell the organism will need from birth to death. Once an organism begins to mature, other cells are required to produce the range of specialized cells needed for a functioning organ or tissue. Such cells are known as stem cells. As technology has developed and the micro-dissection of tissue became possible, it was evident that most, if not all, organs and tissues in the body have their own stem cells which are capable of dividing and differentiating into mature functional cells. In some ways, stem cells can be imagined as blank canvases with many hidden cellular pictures already imprinted on them. Different combinations of proteins (growth factors) or other stimuli such as fats or sugars that touch the cell can stimulate division and allow the daughter cells to take on changed characteristics. In this chapter we look briefly at the types of stem cell and where they come from.

In simple terms, there are two major types of stem cells—embryonic and adult (Figure 16). Embryonic stem cells have different biological properties to the adult stem cells that are found close to and after birth, hatching, or germination. The archetype of stem cells in development is the zygote, produced by the fusion of egg and sperm, which has the complete 'potency' to generate every tissue and cell type in the body (a property termed

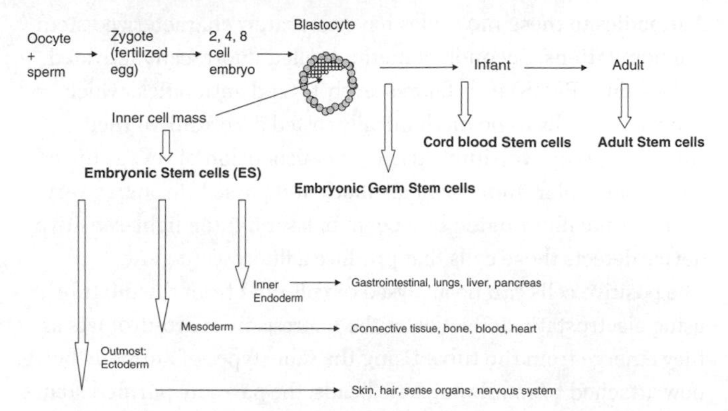

16. The origins of embryonic and adult stem cells

'totipotent'). As the embryo grows, 'pluripotent' stem cells appear, which are limited in their potency. These cells divide and differentiate into the main classes of ES cells (known as the germ layers in mammals) and then, in turn, develop into organs and tissues. From embryo to adult, our cellular growth relies on: 'multipotent' cells that give rise, after division and differentiation, to cell types belonging to a related cell family; 'oligopotent' stem cells that are more limited still, producing just a handful of closely related cells, for example myeloid blood cells; and 'unipotent' stem cells that will give rise to only one cell type, for example muscle cells. The term progenitor or progenitor stem cell is often used to describe those cells still rapidly dividing but not as yet fully differentiated. Adult progenitor stem cells repair tissues by producing specific cell types needed to maintain normal turnover of regenerative organs.

Before we continue the story of stem cells, some developments in scientific technology need to be described as they have greatly accelerated our knowledge of this area. First, there was the serendipitous discovery of the presence of specific proteins and sugar molecules on the surface of various types of stem cells.

Antibodies to these molecules have accurately characterized stem cell populations. Secondly, a method called fluorescent activated cell sorting (FACS) uses fluorescently tagged antibodies, which allow living cells to be mechanically sorted according to their surface proteins. In simple terms, the suspension of cells is mixed with a particular fluorescent antibody and passed through a very narrow tube illuminated by a beam of laser light; a light-sensitive device detects those cells that produce a fluorescent glow. The positive cells can be analysed or collected from the mixture using electrostatic deflection of the fluorescent micro-droplets as they emerge from the tube. Using the same types of antibody but now attached to small magnetic beads, the partially purified stem cells can be isolated in sufficient number for clinical use. Cell populations can also be followed after transfusion into a recipient by attaching a small green fluorescent protein (GFP) gene to one of its own protein genes. Mutations of this jellyfish protein gene produce many different fluorescent colours (responding to light of specific wavelengths for accurate analysis). Whenever this gene is expressed, its protein fluoresces inside the cells. Similar 'bioluminescence' technology has been used to see individual cells within the living animal, often using the sensitive cameras originally developed for satellite surveillance.

Historical perspective

Ignored for two millennia, Aristotle, in his book *On the Generation of Animals*, first proposed the theory of epigenesis in biology, suggesting that development of a plant or animal from an egg or spore follows a sequence of steps in which the organism changes and the various organs form. Though this theory now seems obvious in the genetic age, it was not given much credence because of the dominance of creationist and preformationist theories of life's origins for many centuries. In 1795 the embryologist Caspar Friedrich Wolff famously refuted preformationism in favour of epigenesis. An extended and controversial debate by biologists finally led epigenesis to eclipse the long-established preformationist view. Visual understanding of cell populations continued with improvements in microscopy. At the turn of the 20th century, Ernst Neumann described the cells in the bone marrow and stated that

> The different forms of all blood cells happening in the blood, the lymph-organs and in the bone marrow are all descendants of the 'great-lymphocytic' stem cell. In which way this stem cell completes itself again and again, whether exclusively by a mitotic division or also from other cells.

This was probably the first use of the term stem cell.

Embryonic stem cells

Stem cells found in the embryo can give rise to a large number of cells to generate all 200 of the cell types in the human body. In humans, thc inner cell mass (see Figure 17) of the early embryo has 50–150 cells of three primary types. In 1981, Martin Evans and Matthew Kaufman described a new technique for the culture of mouse embryos and the derivation of cultured embryonic cell lines. Later that year, Gail Martin first used the term 'embryonic stem cell' to describe these cell lines. Eight years later, James Thomson isolated a group of cells from the inner cell mass of the early human embryo and established the first embryonic pluripotent stem cell lines in culture. The current source of many human embryonic stem cells arises from *in vitro* fertilization (IVF) procedures.

While sharing many similar biological properties, mouse and human embryonic stem cells (ES cells) require different environments for their sustained growth without differentiation in plastic flasks. For example, mouse ES cells grow on a layer of gelatin and only need the addition of the protein growth factor LIF, whereas similar human cell lines require a feeder layer of live mouse fibroblast cells and another growth factor (human fibroblast growth factor). Without optimal growth conditions, the cells stop dividing and rapidly differentiate. The growth of various ES cell lines has now been fine tuned and there is an increasing understanding of the genes involved in maintaining stem cell characteristics. The maintenance of pluripotency requires a regulatory network that ensures the suppression of genes that lead to differentiation. The default situation is to limit division and to differentiate. Nature has perhaps evolved an almost foolproof mechanism to protect the organism from the dangers of ES cells that evade the normal controls on growth.

Germ-line stem cells generate sperm or eggs (haploid gametes) which have half the normal number of chromosomes, and transmit genetic information from one generation to the next. Such cells are easily identified, retrieved, and manipulated in the fruit fly. In these flies, eggs develop on a string (or ovariole) within the fly ovary. At one end, a small number of germ-line stem cells move along at a predictable rate and differentiate into eggs within eight days. The stem cells are surrounded by three differentiated cell types—terminal filament cells, cap cells, and inner sheath cells—which help make up an anatomically simple tubular structure (germarium). The cells at the tip of the ovariole are organized into a niche that maintains and controls the germ-line stem cells. A special cell–cell junction is formed between the stem and cap cells. These junctions hold a germ-line stem cell at the anterior and prevent it from moving away where it might receive differentiation cues. A special signal protein is needed to maintain this junction and control the rate of germ-line stem cell division.

In plants, all stem cells are totipotent and are able to divide and differentiate into all the cell types needed to produce the entire organism. Totipotent was a term introduced by the Austrian botanist Gottlieb Haberlandt to describe a property known to all gardeners who have for millennia placed small parts of a plant such as leaves, stems, and roots into soil and water to procreate their precious stocks through cuttings. The first growth of an individual plant cell back into an entire plant (a carrot) was performed by Fred Steward in the late 1950s. Mouse-eared cress, *Arabidopsis thaliana*, is a much-studied small flowering plant useful for genetic studies. The growing tip is a completely undifferentiated tissue found in the buds that continues to make leaves, flowers, and branches throughout a plant's life. The 30–40 stem cells that reside in this complex structure are surrounded by millions of differentiating cells, making this a complex model to study. Most geneticists have resorted to using root tips, which are less complex. By using a mutant that produces a high number of accessible shoot meristems (embryonic tissue), a gene expression map has been made of the meristem that has allowed the characterization and subsequent fluorescent marking of these elusive stem cells.

Adult stem cells

Historically, our understanding of stem cells began with observations of how adult stem cells generate vast numbers of mature daughter cells. Ernst Neumann's far-sighted ideas in the area of blood production from the bone marrow demanded the physical culture of stem cells for the completion of the proof. This proof was not to arrive until 60 years later when researchers discovered various parts of this complex puzzle, showing that stem cells in the bone marrow could self renew as well as dividing and providing all the various mature cells in blood.

In 1961, Ernest McCulloch and James Till devised a series of experiments that involved injecting bone marrow cells into the tail veins of mice prevented from producing their own stem cells by a lethal dose of X-rays. Visible nodules were observed to grow in the spleens of the mice and these were in proportion to the number of bone marrow cells injected. McCulloch and Till speculated that each nodule arose from a single marrow cell, probably a stem cell. This animal-based measure of bone marrow stem cells provided the major tool for measuring stem cell numbers for the next 30 years. In the early 1970s, Mike Dexter showed that it was possible to grow primitive bone marrow cells for many weeks in laboratory culture flasks if they had a stromal cell feeder layer (this is a varied collection of non-blood cell types present in the bone marrow).

In the decades that followed many more tissue-specific adult stem cells have been identified. These stem cells are all capable of division and differentiation into the component cell types of their respective tissue. The observed growth of new neurons in rats has suggested the existence of stem cells in the adult brain. This interesting observation was contrary to previous ideas that brain cells lasted a lifetime. Since then, stem cells have been demonstrated in the brains of adult mice, songbirds, and primates including humans. The growth of new neuronal cells (known as neurogenesis) is restricted to two locations in the brain. Neuronal stem cells can be grown in the laboratory as floating cell aggregates (neurospheres) that contain a large number of stem cells. By changing the culture conditions, they can be differentiated into both neurons (electrically excitable cells that process and transmit information) and glia (cells that feed and protect the brain's neurons).

Other organs have stem cell populations that supply the mature cells required either continuously or at specific stages of development. Examples include breast stem cells that are the source of cells for the mammary gland during puberty. They have been isolated from both human and mouse tissue and, in culture, differentiate into luminal epithelial (the inner layer of potentially milk-producing cells) and myoepithelial cells (the outer layer) as well as having the ability to regenerate the entire organ in the mouse. Human olfactory adult stem cells can be harvested from mucosa cells in the lining of the nose. Rather like ES cells, they differentiate into a wide range of cell types and are seen as a potential therapeutic source because of the ease of harvesting, especially in older people. Hair follicles contain different types of stem cell and they can give rise to neurons (nerve cells), Schwann cells (which contribute to the myelin sheath), myofibroblasts (a cross between fibroblast and smooth muscle cells important in wound healing), chrondrocytes (cells that make and maintain cartilage), and melanocytes (melanin-producing cells that give you a sun tan). Basal cells make up about 30% of the epithelium of the lung and, in humans, are present throughout the airways. They are relatively undifferentiated and probably act as a stem cell for this tissue. Further understanding of their biological properties may lead to their use in lung regeneration. The process of assembling endothelial cells into the blood vessel lining mainly occurs during embryonic development. Initially, it was thought that these cells were derived from endothelial progenitor stem cells early in development, but in the 1990s putative adult endothelial stem cells were identified in adult mouse blood. Recent work suggests that adult endothelial progenitor cells are important in the production of blood vessels, especially when new blood vessel growth is required (and generated) by a growing tumour. Endothelial progenitor stem cells, like blood stem cells, are recruited into the blood stream by growth factors before homing to the tumour site. Destruction of these cells within the bone marrow can reduce the growth rate of a tumour, as no tumour can continue to grow larger than two millimetres in diameter without a blood supply.

Stem cell properties

Adult stem cell numbers remain as a constant small pool of cells within any given tissue. Given their propensity to divide, any escape from control could be fatal to the animal or plant. On the other hand, each cell needs to renew itself as well as supplying progenitor daughter cells capable of further rapid division and differentiation to generate the millions of cells needed for a functional organ or tissue. To explain this observation in bone marrow, Ray Schofield proposed the hypothesis that stem cells exist in a specialized site or niche. This niche is composed of a group of cells dedicated to the provision of a microenvironment

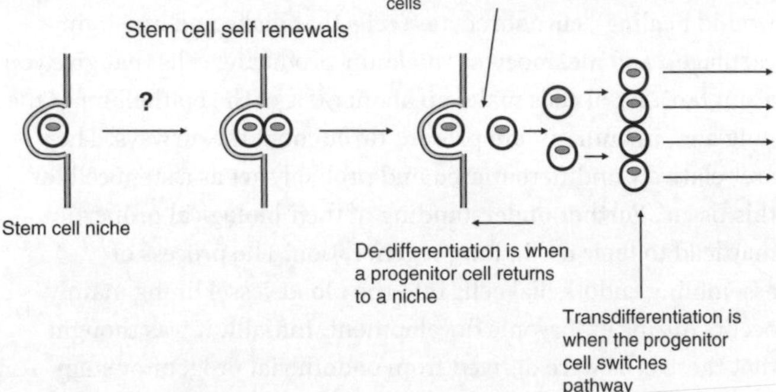

17. A simple model of the stem cell niche. The cell in the niche is dormant until awakened by stimuli not yet fully understood. Following division, the niche accepts one daughter cell while the other, now a progenitor stem cell, divides many times and changes into millions of fully differentiated cells

for the maintenance of a single stem cell (Figure 17). Niches function as a 'base camps' in which stem cells are physically retained, acting as a lasting reservoir for tissue regeneration. By regulating the balance between self-renewal and differentiation, the niche plays an essential role in controlling stem cell fate and maintaining stem cell numbers. In most cases, adult stem cells retained in the niche sites are dormant until they receive signals from the microenvironment that stimulate stem cell division. What constitutes this signal still remains unclear. One daughter cell remains in the niche site as a stem cell while the other, which no longer fits in the niche, leaves and progresses through rapid division and differentiation to form mature cells. If the microenvironment is further stimulated, such as by applying a growth promoting protein, the rate of this process can be greatly accelerated. The niche model has been shown across a variety of stem cell studies in other systems such as in the ovaries of flies, plants, and in the colonic crypts of mammals. Whether this simple model is universal to all stem cells remains unclear.

Adult stem cells have the characteristic of plasticity or transdifferentiation which, in very simple terms, means that one stem cell type can, under different conditions, turn into another cell type. For example, in mice and men, embryonic, bone marrow, adult liver progenitor, and other stem cells can all produce mature cells of the liver. These changes can be performed in the laboratory by introducing growth factor proteins to the stem cells or by transplanting cells into the liver where they are able to repopulate and, in some cases, even improve liver function.

In the embryo there are three primary germ layer cell types: ectoderm (giving rise to the nervous system, tooth enamel, hair, and keratinocytes of the skin), endoderm (developing into the guts, respiratory system, and bladder), and mesoderm (responsible for bone, muscle, connective tissue, the middle layer of the skin, liver, and bone marrow). Surprisingly, all three of these cell types have been found to produce mature cells normally derived from a different lineage.

What is not clear is how this transdifferentiation of stem cells works. The fate of the stem cell is defined in part by its genetic profile at division but also by the external signals it receives. If hit by conflicting external signals, the cell can switch its genetic profile and change into a different cell type. Because they are relatively small and structurally indistinct, the visual identification of true adult stem cells has always proved difficult. It has been suggested that plasticity could be explained by the presence of two or more populations of stem cells within the tissue. For example, adult stem cells give rise to mature cell types, while a smaller number of germ stem cells are still able to produce all cell types. Another model that tries to explain the balance between self renewal of stem cells and those differentiating in the bone marrow, hair follicles, and the gut, proposes that a stem cell can exist in two different states. One is dormant, a stem cell retaining a full developmental potential, whereas the other stem cells are active, capable of producing large numbers of differentiated cells. The balance between inactive and active stem cells is controlled by the levels of various developmental signalling proteins which were first identified in fruit flies but now known to be vital to all animal cells.

Normal metabolic activities—especially DNA replication, when combined with environmental factors such as carcinogenic chemicals, UV light, and radiation—cause DNA damage. It is calculated that as many as one million individual molecular lesions per cell per day can occur. The cell has a collection of processes by which these damaged DNA components are identified and repaired. Most of these DNA changes are harmless, and some probably contribute to why we are all different. Some cause serious structural damage to the DNA molecule and alter the ability of the cell to survive. Examples of these lethal changes include the chemical cross-linking or breakage of the DNA strands. Failure of the normal processes of DNA repair is recognized in the cell by the protein p53 (mentioned in Chapter 3). The cell then either slips into a permanent state of dormancy known as senescence or dies by apoptosis. In the absence of a p53 response, the damaged cell may begin to undergo uncontrolled cell division resulting in a cancer.

All animal cells have an internal ageing clock. Every chromosome has protective structures at each end called telomeres, which are made up of repeats of the DNA sequence TTAGGG. Telomeres protect against the ends of chromosomes fusing, preventing the formation of ring chromosomes. Each time the cell divides, one or two copies of this sequence are not replaced. Eventually, after many divisions, this protection at telomeres begins to run out and the chromosome ends begin to 'fray', at which point division ceases. Embryonic stem cells avoid this limitation on the numbers of 'permitted' divisions by producing an enzyme called telomerase that repairs the damage and allows for the multiple divisions required in early development. In adult tissue, cells that need to divide continuously (for example immune cells and organ-specific stem cells) also have high levels of telomerase, whereas most other cell types express it at low levels. High levels of telomerase are often found in rapidly dividing tumour cells.

The stem cells at the growing tips of plants remain active throughout life, which may be centuries in some trees, during which time they are constantly exposed to environmental hazards that cause DNA damage and mutations. All plants have mechanisms that respond with great sensitivity to DNA damage, whether from radiation or cytotoxic chemicals, and signal an early cell death. Similar observations have been made in mouse bone marrow stem cells. The evolution of this stringent way of retaining their genomic integrity in these crucial cell populations is distinct from normal programmed cell death. On the other hand, there is an observed decline in the numbers of adult stem cells in many mammalian tissues with time, probably resulting from gradual low-level DNA damage. This affects their ability to divide with age by pushing cells into a dormant state. The number of stem cell niches may also decrease. We grow old because our stem cells age as a result of mechanisms that in our youth suppress the growth of cancer cells.

Cancer stem cells

The idea of cancer stem cells has slowly increased in credibility over the last ten years. The leukaemia stem cell was the first such cell to be described but it has been suggested that cancers of the brain, colon, ovary, pancreas, and prostate may also have stem cells. The origins of these cells are controversial. The concept is based on the notion that, as with normal stem cells, a pool of malignant cells has the ability to indefinitely self renew and re-initiate tumours in distant locations. As with normal stem cells, these cells only divide slowly (in other words self renew), which makes them more resistant to anti-cancer drugs that usually kill rapidly dividing cells, while their differentiated daughter cells remain sensitive. One explanation of leukaemia stem cells is that the genetic damage leading to cancer takes place in a newly differentiated cell followed by a de-differentiation event triggered by the stem cell niche, resulting in a very limited but immortal supply of leukaemic stem cells. In solid tumours it is extremely difficult to pinpoint the precise origin of the putative cancer stem cell as the tumour has a heterogeneous population of mutant cells. Among these cells there may be several types of stem cells, one optimal to the specific environment and several less successful lines. The latter cells can become more successful in some environments, allowing the tumour to further adapt to changes in its environment. Ongoing stem cell research may have repercussions on new cancer treatment.

Chapter 7
Cellular therapy

Most diseases have cellular origins. The majority can be traced to simple biochemical imbalances that can often be corrected by drugs. The cells return to a near-normal state and the patient feels better. Some complex diseases start with gene changes that prevent a particular protein being expressed or change the protein structure in such a way that its function is altered or ineffective. Even with the checks and balances of internal repair or cell suicide, some gene changes escape and, when involved in the fundamental cell processes, they can have a devastating effect on cell growth or function. Increased growth caused by gene changes leads to the slippery slope of cancer where these 'out of control' cells further change their genetics and develop the ability for limitless division and, in some cases, gain the ability to invade other tissues. In simple terms, treating such diseases requires that the errant cells are removed, corrected, or even better replaced. Many other diseases are not directly life threatening but cannot be corrected by drug treatment because of their complex biochemical nature. The ideal solution would be to treat the disease at its root by replacing the faulty cells. This process is called cellular therapy and can be divided into different forms. The most familiar is the transplantation of mature functional cells, as in a blood transfusion, and stem cells as in bone marrow transplants. We still await a standard therapy for the introduction of modified human or animal cells that replace a required substance, such as insulin-producing cells for diabetics.

Many quote the Swiss physician and alchemist Philippus Aureolus Paracelsus as the first person to describe the concept of cell therapy in his book *Der grossen Wundartzney* ('Great Surgery Book') published in 1536 in which he wrote: 'the heart heals the heart, lung heals the lung, spleen heals the spleen; like cures like'. This remark came from his theory that eating healthy animal organs would rebuild and revitalize the particular ageing or faulty organ—more of a nutritionist doctrine than one of a modern cellular therapist. As long ago as 1667, Jean-Baptiste Denis, working in the royal laboratory of Louis XIV, attempted to transfuse blood from a calf into a mentally ill patient. The first recorded non-blood transfusion occurred in 1912 when German physicians tried to treat children suffering from under-active thyroid with transplanted thyroid cells. In 1931, a Swiss clinician Paul Niehans became the 'father of cell therapy' by accident when he introduced minced bovine thyroid tissue into a rapidly deteriorating patient who had experienced severe damage to his thyroid glands during surgery. The patient recovered and lived for a further 30 years. Niehans became renowned for his cellular treatments, and his patients included members of various Royal families, Pope Pius XII, politicians, and famous film stars. His name and some treatments, especially skin and beauty processes, live on at the Paul Niehans Clinic founded in Switzerland by his daughter. The 20th century is littered with stories of so-called cellular therapies. John Brinkley, known as the 'goat gland doctor', reportedly did 16,000 operations in which he implanted men with tissue from the testicles of young goats, asserting that this procedure was effective against impotence and could cure conditions ranging from acne to insanity. His licence was revoked on the grounds of immorality and unprofessional conduct. James Wilson promoted the use of bovine connective tissue cells. He claimed that such cellular preparations taken by mouth 'have the ability to migrate to any tissue in need of repair and, once at the site, take on the characteristics of the healthy cell it associates with'. Other attempts along similar lines caused hundreds of deaths by violent immune reactions to bacterial or viral infections including two men who died of gas gangrene following injections of foetal sheep cells. Other, even more questionable embryonic cell transplant therapies were pioneered by Niehans in Switzerland. In the 1970s, his student Wolfram Kuhnau set up a clinic in Mexico, where there were no regulations, using embryonic blue shark cells purchased off local fishermen and with a dubious number of live cells, to 'treat' patients for a wide range of diseases. As a result of this 'quackery' and the bad press it received, it is no wonder that there was a high degree of scepticism about cell-based therapies as we approached the 21st century.

Transplantation of blood cells

Blood transfusion using animal blood started in the late 15th century but was unsuccessful because of blood group incompatibility and infections. Such was the death toll that transfusion was banned throughout Europe for over 150 years. Around the turn of the 20th century, Karl Landsteiner discovered human blood groups (A, B, AB, O), for which he received the Nobel Prize in 1930. Mixing blood from individuals with incompatible groups leads to blood clumping or agglutination. Landsteiner discovered that this process was due to an immunological reaction between dissimilar blood groups. The classification of blood groups is based on the presence or absence of inherited antigens, including proteins, carbohydrates, and lipids, on the surface of the red blood cells. If incompatible blood is introduced, antibodies in the recipient's blood plasma attack the new red blood cells and destroy them in a reaction called haemolysis, resulting in renal failure and circulatory shock. Blood transfusion is now a routine procedure and most blood donations are fractionated into their components. These include red blood cells, platelets, white blood cells, plasma, and various proteins such as antibodies and clotting factors (see Chapter 5). Selected components allow for specificity of treatment, reduction of side effects, and efficient use of a unit of blood. For example, platelets can help restore the blood's clotting ability when given to people with too few platelets (thrombocytopenia), a condition leading to severe and spontaneous bleeding and which is sometimes a side effect of chemotherapy. While the transfusion of white cells is rare for treating infections, it has been used in therapy when the cells have been genetically manipulated to elicit any anti-tumour cell activity.

In the 1970s, Edward Donnall Thomas demonstrated that bone marrow cells infused intravenously can repopulate the bone marrow and produce normal blood cells, work for which he received the Nobel Prize in 1990. The active cells in these repopulating preparations are adult blood stem cells. Cytotoxic drugs required for cancer treatment destroy all dividing cells and consequently do not discriminate between tumour cells and blood stem cells. Replacing the patient's stem cells following chemotherapy replenishes the bone marrow cell population and promotes the resumption of normal blood cell production. Bone marrow transplantation has been used in a wide range of cancer therapies including leukaemia and lymphoma to repair bone marrow damage caused by chemotherapy. Bone marrow transplants have also been successful in treating cases of anaemia and other diseases where stem cells are damaged or absent.

Autologous blood stem cell therapy uses the patient's own stem cells, which are removed and purified prior to treatment. The advantages of autologous transplants are that they are relatively free of infection and the recovery of immune function is more rapid and increased. Autologous transplants are not always possible and so other sources of stem cells are needed. Allogeneic blood stem cell transplants involve a healthy donor whose tissues must be a close tissue match to those of the patient. The closer the donor (often a relative) is genetically to the patient, the better the match of specific cell-surface proteins, called the histocompatibility complex. Even a single DNA base pair difference resulting in a changed amino acid sequence of one of the five histocompatibility proteins will result in a mismatch. Leading bone marrow transplant centres are able to sequence the DNA of all five histocompatibility genes to check for compatibility. Only identical twins give perfectly matched stem cells but unrelated donors need as many matches as possible. Mismatching increases the risk of graft rejection or graft versus host disease. Graft rejection occurs when the body rejects the new transplanted cells while graft versus host disease occurs when the new cells reject the body. Both effects result in an immunological reaction that can be fatal.

In early treatments, bone marrow was taken from a large bone of the donor, typically the pelvis, through a needle that reaches the centre of the bone. The technique is referred to as a bone marrow harvest and is performed under general anaesthesia with all the requirements of hospitalizing the donor. Nowadays stem cells can be sourced from circulating blood. This method was developed from an observation that circulating stem cells increased dramatically following injections of a blood protein growth factor. Donors are given growth factor and their stem cells are collected by a mechanical separator, after which the red cells are returned to the donor.

It is also possible to isolate useful amounts of blood stem cells from amniotic fluid and the umbilical cord. Cord blood has a high concentration of stem cells, but only enough for blood stem cell transplants in young children. Using combinations of growth factors, it is possible to amplify the numbers of stem cells in the umbilical cord allowing possible use in adult transplants. Umbilical stem cells generally produce low levels of graft versus host disease. Storage of one's own umbilical cord stem cells for possible future use in adult life has developed into an expanding business for tissue storage companies.

Organ stem cell therapy

In the early 1960s, mouse embryonic carcinoma cell lines having stem cell characteristics were derived from a teratocarcinoma. This cancer tissue is made up of germ cells that are derived from the ovary directly or indirectly due to birth defects resulting from errors during embryo development. The use of these cells was hindered by problems of genetic mutations and genome instability. The isolation of ES cell lines from normal embryos (see Chapter 6) overcame these defects and began a new area of research, exploring the possibilities of isolating and manipulating human embryonic cells for possible use as stem cell treatments in adults.

Mouse ES cells have been used extensively in the generation of transgenic mice. These genetically modified mice are extremely useful, as they act as a model to study the functions of individual genes in a species close to humans. Transgenic mice are created by transferring a gene (for example a human cancer-related gene) into cultured ES cells. Alternatively, individual genes can be 'knocked out' by introducing a foreign gene—often a drug resistance protein—into a target gene. Unfortunately these gene manipulations are notoriously inefficient, and consequently the treated cells need to be grown in culture and selected for the correct gene change. The selected cells are then micro-injected into the inner cell mass of a normal embryo before implantation into the uterus of a foster mother mouse. The offspring have a copy of the introduced gene present in one of their paired chromosomes (i.e. they are heterozygous) and can be further bred to gain homozygous strains in which the new gene (or knocked out gene) has a copy in both matched chromosomes, so the effect will be apparent in every cell in the animal. Whether the gene is expressed in a particular tissue or cell type will depend on its genomic environment, something that can usually be directed at the ES cell level by careful molecular selection. There are now thousands of transgenic mice strains each with a specific gene alteration and they have greatly enhanced our understanding of many complex biological processes. When an error or the over expression of a particular gene is the major cause of a disease, transgenic animals can act as a model for developing new drugs or pharmaceutical interventions. Even when these genetic manipulations introduce changes that result in aborted foetuses, cell lines can be 'rescued' and grown in culture for useful research.

There are a number of possible sources of human ES cells: (i) cadaver foetal tissue, (ii) embryos remaining after infertility treatments, (iii) embryos made solely for research purposes using IVF, and (iv) embryos made using nuclear transfer into eggs (the technique used to create Dolly the sheep). All human ES cell sources raise ethical and religious questions. Country by country, political regulation usually reflects the predominant ethical and religious views, with many countries completely prohibiting all work on embryos, others having strict regulations, while some have few, if any, restrictions. Opponents of human embryo stem cell research believe that a human life begins as soon as an egg is fertilized. The destruction of an embryo is deemed morally as murder. They also argue that ES cell technologies are a first step to reproductive cloning, which fundamentally violates and devalues the sanctity of life. Proponents argue that in the natural reproductive process, human eggs are often fertilized but fail to implant in the uterus. A fertilized egg, while it may be capable of forming human life, cannot be considered a human being until it has been successfully implanted in a woman's uterus. The methods required for IVF routinely create more human embryos than are needed over the course of a fertility treatment, leaving excess embryos that are often simply discarded and it is morally permissible to use such embryos for potentially life-saving biomedical research. ES cell lines from early 1990s experiments have been allowed limited use in clinical treatments.

One example of using ES cells to treat human injury comes from the approval in 2009 for the first phase of clinical trials for transplantation of human stem cells of the brain and spinal cord (oligodendrocyte progenitor cells), derived in culture from human ES cells, into patients with injured spinal cords. The first patient was treated in October 2010 by Hans Keirstead's team at the University of California, Irvine and sponsored by the biotechnology company Geron. Experiments in rats had previously shown that there had been improvement in the recovery of movement in animals with spinal cord injuries after a seven-day delayed transplantation of human ES cells that had been forced into oligodendrocyte lineage in culture. This new ongoing study to treat paraplegics will last for at least five years. While it is not expected that this treatment will completely cure the injury, it is hoped that it will show sufficient repair of the nerve cells to be worthwhile. This is a pioneering study promising a bright future for the use of ES cells for treating spinal injuries.

The ethical, religious, and political constraints of isolating embryonic cells for human therapy has driven the research dream, and now the reality of inducing or de-differentiating adult cells into embryo-like pluripotent cells that could be used in cellular therapy. In 2006, Shinya Yamanaka produced induced pluripotent cells from mouse fibroblast cells by forcing the expression of several genes. Retroviruses (a DNA virus family that inserts its sequence into the host DNA) were used to induce the expression of genes identified as important for maintaining ES cells in the adult fibroblast cells. Early attempts did not prove totally successful as the expression of some viral genes caused cancer after transplantation into the mouse embryo. This problem was overcome in 2008 by using an adenovirus to introduce genes. Unlike retroviruses, this virus does not incorporate its own genes into the host cell genome. Several groups mirrored these mouse experiments in adult human cells. In the following year, Sheng Ding and colleagues found it was possible to change somatic cells into pluripotent cells without oncogene insertion but using repeated treatment with two small chemically synthesized proteins. The manipulation of an individual's adult cells to produce embryonic-like cells capable of repairing their organs and tissues is predicted to be a major clinical treatment of the future.

The potential of using stem cells to repair diseased and damaged organs without the risk of organ rejection or other side effects is an endpoint that is driving much research. Stem cell therapies exist but, so far, only as experimental medical treatments. The main areas of progress are in the treating of cardiac and muscle damage, diabetes, liver, Parkinson's disease, and Huntingdon's disease.

Cardiovascular disease—including hypertension, coronary heart disease, stroke, and congestive heart failure—is the major cause of death in many countries around the world. When deprived of oxygen, cardiac muscle cells (cardiomyocytes) die and this triggers the formation of scar tissue, the overload of blood flow and pressure, and the over-stretching of viable cardiac cells leading to heart failure and death. Using mouse, rat, and pig models, various types of stem cells including embryonic, mesenchymal, endothelial, and naturally occurring heart stem cells have been shown to regenerate damaged heart tissues. A few studies in humans, usually undergoing heart surgery, have shown that stem cells introduced directly to the heart or transfused into the blood circulation have given improved cardiofunction and induced the formation of new capillaries.

Muscular dystrophy is a group of genetic disorders in males that causes the muscles to weaken with time and eventually leads to premature death. The condition is caused by changes in the protein dystrophin which normally maintains the integrity of muscle. Using mouse and dog models, stem cells called mesoangioblasts (programmed to differentiate into muscle cells) with a corrected dystrophin gene have been transplanted. Normal dystrophin levels and muscle strength were regained in four out of six dogs, suggesting that cellular therapy may be the way ahead for treatment of this genetic disease.

In 2008, ES cells were coaxed into developing into immature beta cells capable of producing insulin, which were able to reverse diabetes in mice. In the same year, adult human skin cells, first induced into pluripotent cells, were also reprogrammed to produce insulin. A more exciting aspect was that these cells secreted insulin in response to glucose (as they do normally in the cells within the pancreas). It is only a matter of time before transplantation of insulin-producing cells will replace the need for a lifetime spent injecting insulin in people suffering from type 1 diabetes. Similar progress has been made culturing liver cells from induced adult stem cells in mice, making liver regeneration a distinct possibility.

The gradual loss of dopamine-producing nerve cells in certain areas of the brain results in Parkinson's disease. Early cellular treatments relied on transplanted foetal brain tissue. A few individuals showed a marked improvement and provided a general proof of principle. Apart from the contentious tissue source, these clinical trials did highlight several issues including the need for large quantities of pure cells and that the foetal transplants became affected themselves by transmission of Parkinson's disease. In 2008, induced progenitor stem cells were created from skin fibroblasts of mice and then further differentiated into neuronal precursor cells which (following transplantation) integrated into the surrounding brain. This approach has been further extended in an animal model which mimics Parkinson's disease by killing off the normal dopamine-producing cells with a toxin. Transplantation of induced stem cells gave a marked improvement compared to non-treated animals. The major challenge is now to understand how Parkinson's disease develops and to use stem cell models to develop new drug treatments.

Huntington's disease is a neurodegenerative disease that is characterized by the death of neuron cells of the brain that produce a neurotransmitter chemical (gamma aminobutyric acid). Clinical treatment using stem cells for Huntington's has mirrored that of Parkinson's disease. Foetal tissue and induced adult stem cells grafted into the adult brain increase brain activity, including motor and cognitive functions. Stem cells may not be the 'magic bullet' for this disease but could form a major part of its effective management.

Mesenchymal stem cells, found in the bone marrow, can be induced to form cartilage, bone, tendon and ligaments, muscle, skin, fat, and nerve cells. They are easy to isolate from small quantities of bone marrow and are readily grown to the amounts needed for transplantation. Frozen stocks are viable and could be used for 'off the shelf' therapies. They are also found in dental pulp, the soft tissue inside teeth. Wisdom teeth are a particularly rich source of stem cells and this opens up the interesting possibility of visiting the dentist for a stem cell transplant. Embryonic mesenchymal stem cells have been transplanted into the jaw bone of adult mice. Some of these 'tooth germs' grew into fully functional hard teeth capable of responding to pain. Whether adult tooth-derived stem cells can be persuaded to produce regenerated teeth remains uncertain but an Indian biotechnology company has set up the world's first dental stem cell bank.

Chapter 8

The future of cell research

The whole cell is considerably more than the sum of the working parts. The same can also be said about the genome, where the identification of the blueprint for individual molecular components of the cell is undertaken in the expectation that a rigorous characterization of all of the parts separately will lead to an understanding of the whole. This is a system of investigation called reductionism, which has been a dominant philosophy in biological investigation for decades. However, just to identify the molecular parts of the puzzle is not going to tell us how the whole works if we do not understand the rules for their assembly. This requires the development of approaches to investigate 'systems biology' or 'biocomplexity', and represents a paradigm shift (i.e. 'thinking outside the box') in biological research, wherein the challenge is to understand the collective interactions of multiple molecular processes, not only within the cell itself, but also at the tissue, organ, and organism level. The bottom line is 'do the molecules drive the cell to drive the organism, or does the organism drive the cell and its molecules?' In reality, such interactions lie somewhere between the cell responding to its immediate environment, balanced against the controls of gene expression.

There are many new exciting areas of research into cells. Some involve the understanding of how cellular behaviour can change so dramatically after subtle changes at the level of genes and proteins. Other approaches use cells to cure a disease, for extraction of metals, the breakdown of petroleum products, or to harness light to produce biofuels. Equally fascinating are our efforts to create synthetic life and understand the biology behind ageing. In this chapter we briefly examine some of these future directions and how they might evolve.

Systems biology

We now know the complete DNA sequence of just a few humans. We also have a rapidly increasing understanding of the biochemical mechanisms involved in the day-to-day existence of the cell and how it divides and differentiates. In the past ten years, advances in molecular technology have allowed us to induce and monitor changes in thousands of genes, their accompanying RNA signals, and protein production. These changes can now be measured more or less at the same time and even within a single cell. The knowledge coming out of this collection of technologies has developed into its own subject, known as systems biology. It has allowed us to see millions of more subtle interactions between the different components of the cell. Early experiments monitored the changes that occur within a cell when it is subjected to a known pharmaceutical drug. For example, in the simplest case, a drug interacts with its target enzyme protein and stops it working. What is apparent now through our ability to analyse thousands of individual genes and their products simultaneously is that a drug also triggers changes in the levels of many other proteins, often seemingly unconnected to the original target enzyme, which may be increased or reduced, often at differing rates. This may account for some drug side effects but also allows the development of 'cleaner' more specific pharmaceuticals. By further applying this methodology to various biological systems, we are starting to discover the alterations in gene expression and protein levels that take place during various biological processes such as cell division and differentiation. While these experiments themselves take relatively little time to perform, understanding what they mean will take longer as the vast amount of data generated needs to interpreted. Fortunately, analysis of this information has been made possible by powerful computing. Increasingly, cell and molecular biology relies on this *in silico* biology, which is known as bioinformatics, to solve the difficult questions of biological behaviour in terms of DNA and protein sequence.

Leroy Hood, a pioneer of this field, suggested a beautiful example of how systems biology might work in medicine. A patient attends their doctor's surgery, giving a pin prick of blood. What follows is the full biochemical, gene, and protein analysis of the body's function and health state. This information will be instantly computed for all possible diseases, matched with the symptoms, and therapies or further tests suggested all in a matter of minutes. This is personalized medicine, a dream a few years ago, and now only a matter of (uncertain) time. Cancer researchers are already using advanced protein and DNA technology to monitor the extremely small number of cancer cells that can be found in the blood of a patient suffering from a solid tumour (a small number of tumour cells are continuously shed into the blood, passing through the capillaries supplying nutrients to the growing tumour). Pharmaceutical researchers are using these methods to investigate the general efficacy of new anticancer drugs.

Given the dynamic nature of the living cell, to follow the fate of individual proteins during biological processes (for instance during stem cell differentiation) requires the ability to tag one or more proteins and observe them in real time. Previously, molecules such as green fluorescent protein (GFP) would be used as labels but this tag could be many times larger than the molecule of interest, and possibly interfere with its normal activity. Tagging is now possible with minute inorganic spheres (quantum dots) which are so small (see Figure 1) that they pass straight through the cell membrane. Nanotechnology is the science of controlling matter on an atomic or molecular scale, and involves the interaction of materials between 1 and 100 nanometres (DNA strands have a diameter of two nanometres). The massive advantage of working at this scale is the ultra fast speed of the reactions, which are reminiscent of those found in the cell. This growing area of research is being applied to living cells in the molecular analysis of disease. Future applications will include the micro-manipulation of faulty genes, building cellular bio-sensors, and creating DNA computers. Imagine a time when an oral medicine is not just a chalky pill, but will consist of a capsule full of nanorobots programmed to find and reconstruct the DNA of cancer cells or dismantle a life-threatening virus. Cells from patients with an inherited disease could be corrected and affected organs could be restructured by surgical nanorobots.

Synthetic life

All living cells are related to each other by their use of the same genetic code and a small number of highly conserved protein sequences. This suggests that all modern life evolved from a single ancestral living entity. The components of the cell—DNA bases, amino acids, and even small polymers of these—have been created in the laboratory using recreations of the extreme chemical and physical conditions that existed as the new Earth cooled down. We are now in the era of creating entirely synthetic cells from elemental precursors.

Cells need to create copies of their molecules before they can divide, but they also need the software and all the complex protein synthesis machinery to get there. Therefore, the minimum requirements for a cell are: a containing structure; a DNA sequence laid out in the logical order of genes, each coding for a protein that can perform a simple chemical reaction; and, most importantly, the ability to bring these processes together with the information to consistently replicate itself. Fat molecules can self-assemble into primitive membranes, forming spherical structures to protect and concentrate their contents. Artificial ribosomes capable of assisting in protein synthesis, and functional synthesized genomes introduced into cells lacking a nucleus have been created successfully in the laboratory. Bioengineers have recently created a photosynthetic foam containing all the enzymes needed to convert 98% of sunlight into sugar. These biotechnology applications currently mimic natural life processes. Whether we will be able to recreate a complete independent life form is less certain but such work may help us to further define what life is.

Growing limbs and ears

In our own bodies, only the liver is capable of limited regeneration, but chop a limb off a starfish or a salamander, and it will grow a new one. We are starting to understand the molecular signals that are used by these species to regenerate limbs in adult life. Mammals seem to only use this signalling pathway during the growth of the early embryo but it is a pathway that may well have the potential to be reactivated. Following surgical removal, the wings can grow back in embryonic chickens when the production of a protein called wnt is switched on. Frog limb regeneration can also take place later in the life cycle when wnt protein is expressed. Tadpoles have this ability but it is normally lost when they metamorphose into frogs. The expression of wnt signalling protein around an injury is thought to cause a reprogramming or transdifferentiation of mature cells into stem cells capable of producing the cell types needed for the limb. Very young children have been known to re-grow severed fingertips, and so there are intriguing possibilities for human tissue regeneration.

An eternal existence

Can we live forever? Instead of growing old, the 'immortal jellyfish' (*Turritopsis nutricula*) reaches sexual maturity, then reverts back to its juvenile polyp stage (equivalent to embryonic cells). In mature jellyfish (medusa stage), cells from the bell surface and digestive canal organ become transdifferentiated into polyps which then develop into mature jellyfish and so on. This property of bypassing death by reversing its life cycle is, so far, unique in the animal kingdom, allowing a sole jellyfish to live and reproduce indefinitely. Fortunately, most of these jellyfish (they breed every 24 hours) are lost to the general hazards in the sea.

Recent studies have shown that an insulin-like receptor protein in the nematode worm *C. elegans* plays a vital role in controlling lifespan. This gene is important in regulating reproduction, heat tolerance, resistance to the lack of oxygen, and bacterial attack. Mutations in this gene allow the worm to live twice as long (albeit in laboratory conditions). Higher levels of a protein that controls the expression of this insulin receptor are also correlated with longevity. The worm leads the way to examining whether the equivalent genes in mammals are able to prolong life. As we understand more about how cells age, it may be possible to manipulate the genetics and mechanics of this process.

An alternative way of gaining immortality is to create a clone of yourself. The first mammal cloned was Dolly the sheep by way of somatic cell nuclear transfer (introducing the nucleus of an adult cell into an unfertilized egg followed by surrogate motherhood). This has led to the production of a range of 'immortal' cloned animals including beloved household pets. The problem with producing a new animal essentially from an adult nucleus is that it increases the risk of the early onset of age-related diseases including cancer. Human cloning and the growth of the early embryonic body has been suggested as a new and convenient cell source to provide therapeutic stem cells for transplantation. Ethical and political concerns have stopped further work in this area for the moment.

Life on this planet is based on the cell. Single-celled life forms like bacteria, yeasts, and algae have evolved over many millions of years into complex multicellular animals capable (in our own situation) of trying to understand how life itself works. Meticulous observation and profound thinking by 19th-century biologists helped us to understand what the cell is, how every new cell is related to its mother cell through its genes, and how one cell or collection of cells can evolve into new species by adapting to a changing environment under the influence of natural selection. The 20th century has seen the unravelling of the cell's components, the decoding of the vast amounts of information that lies in DNA and proteins, and an initial grasp of the complex communication between cells and the molecular components involved. The 21st century will probably yield ways of using cells, natural or synthetic, to cure disease, regenerate any part of the body, and extend lifespan. Perhaps we can even create cell-based supercomputers. Just as the evolving bacteria in Earth's early history changed the chemistry of the biosphere, similar approaches could be used to reverse the harm we have done in our continuing exploitation and pollution of the planet. Whatever happens in the future, living cells, in some form or other, are likely to survive and adapt to their environment. Whether humans will be around to see this is rather less certain.

Further reading

David S. Goodsell, *The Machinery of Life*, 2nd Edition (New York: Springer-Verlag, 2009). Explores the application of systems biology to individual cells, suggesting that they are controlled by molecular circuits that provide the basis of the properties of all living systems.

Lewis Wolpert, *How We Live and Why We Die: The Secret Lives of Cells* (London: Faber and Faber, 2009) for more general interest, and written for the interested lay reader rather than the undergraduate.

Nick Lane, *Power, Sex, Suicide: Mitochondria and the Meaning of Life* (Oxford: Oxford University Press, 2005). An intriguing view of a particular cell component, the mitochondrion, which suggests that this organelle has been the moving force that has driven cells to their current level of complexity.

Denis Noble, *The Music of Life: Biology Beyond Genes* (Oxford: Oxford University Press, 2006). A pioneer in systems biology, Noble argues the case that the reductionist view that 'genes are everything' needs to be balanced by a systems approach, using the analogy of music and the orchestra.

Rebecca Skloot, *The Immortal Life of Henrietta Lacks* (New York: Random House, 2010). A ten-year study by the author chronicles the life and early death of the source of the first human cell line, together with an account of the scientific, social background and healthcare systems in the USA in the post-war years.

Bruce Alberts, Alexander Johnson, Julian Lewis, Martin Raff, Keith Roberts and Peter Walter, *Molecular Biology of the Cell* (New York: Garland Science, 2008). Classic textbook now in its 5th Edition.

Véronique Kleiner and Christian Sardet, *Exploring the Living Cell DVD* (Meudon, France: CNRS Images, 2006). Nineteen short films from top international institutes, covering cell discovery, cell structure and stem cell biology and ethics.

Websites

Some of the leading science journals allow web access to current material:
Science
http://www.sciencemag.org
Nature
http://www.nature.com
Scientific American
http://www.scientificamerican.com
New Scientist
http://www.newscientist.com
Cell
http://www.cell.com
Journal of Cell Science has free access to short explanations of specific topics in a series called *Cell Science at a Glance*.
http://jcs.biologists.org
The iBioMagazine, published by the American Society for Cell Biology, offers a collection of short (<15 min) talks that highlight the human side of research.
http://www.ibiomagazine.org
Molecular biology animations by John Kyrk.
http://www.johnkyrk.com
Dynamic animations of cell structure, including a graphic representation of Tensegrity (University of Toronto).
http://www.molecularmovies.com/
http://multimedia.mcb.harvard.edu/
Dennis Noble's elegant presentation of the contents of his own short volume *The Music of Life* is an entertaining 45-minute argument against reductionism using an orchestra as an analogue to the cell.
http://www.pulse-project.org/node/25
There are some good YouTube clips of recent discoveries explained by scientists. *Apoptosis* (Walter and Eliza Hall Institute, Melbourne) is worth viewing for its content, especially the sound effects.
http://www.wehi.edu.au/education/wehitv/

Glossary

Apoptosis A type of cell death in which the cells are genetically programmed to die, often in developmental programmes such as the loss of webs between fingers in the human hand.

Archaea Previously classified with bacteria, Archaea are a similar group of primitive single cell organisms that can also live in the most extreme conditions on earth.

Axoneme A central complex of microtubules which supports cell extensions such as the classic 9+2 organization found in cilia and flagellae.

Ciliopathies A group of genetic conditions resulting from defects in cilia, flagellae, or their basal bodies, leading to a variety of different diseases.

Condensation The final packaging of chromosomes at metaphase, prior to separation of chromatids to the daughter cells.

Cytokinesis The last stage of cell division, following nuclear division, when the mother cell cytoplasm is separated into two daughter cells.

Cytoskeleton The combined organization of microtubules, intermediate filaments, and microfilaments, which works together to control cell shape, locomotion, and division.

Dalton The unit of atomic mass, equal to one atom of hydrogen.

Desmosome A specialized region of the cell membrane between adjacent cells in tissue, forming a junction which joins cells together, anchored in the cytoplasm by intermediate filaments.

Diploid Normal amount of DNA in somatic cells, in the form of paired (homologous) chromosomes, in contrast to germline cells (sperm and eggs) which have a single set of chromosomes, and are haploid.

Endosymbiosis The mutually advantageous relationship of previously free-living precursors of mitochondria and chloroplasts with the larger cells they came to live in early in the evolution of eukaryotic cells.

Epithelium Cells that line surfaces in the body. One of the four basic types of tissue that make up the body, along with nervous tissue, muscle tissue, and connective tissue.

Eukaryote A cell with a clearly defined nucleus, enclosed in a membrane, and with other membrane-bound organelles in the cytoplasm.

Exocytosis The release of material from inside the cell, mediated by fusion between vacuolar and plasma membranes, which occurs during secretion of cell products.

Exon A sequence of bases, usually in messenger RNA, ready for translation into a specific protein after the removal of introns during splicing in post transcriptional processing.

Fibroblast A cell found in connective tissue, which secretes collagen and extracellular matrix, and responds to damage by pulling wound margins together.

Gamete A mature sexually reproductive cell (egg or sperm in animals, pollen and ovum in plants) which fuse to from the zygote, the diploid precursor, and ultimate stem cell of every organism.

Gene The unit of heredity, carrying the instructions for production of a specific protein coded in the form of a series of bases in a sequence in DNA.

Haploid *see* Diploid.

Intron A sequence of bases removed from a gene during RNA splicing (*see* Exon).

Laminopathies A series of rare genetic conditions resulting from mutations in proteins of the nuclear lamina, e.g. Emery-Dreyfus muscular dystrophy.

Lipogenesis The formation of fat from the metabolism of simple sugars.

Millimetres, microns, and nanometres. A millimetre is equal to one thousand microns, and a micron equals one thousand nanometres. A whole eukaryote cell may be 10 microns in diameter, part of that cell, e.g. a microtubule is 25 nanometres in diameter. The diameter of human hair varies between 50 and 100 microns.

Nucleoskeleton An intranuclear network of filamentous proteins, anchored at the nuclear envelope which maintains nuclear integrity and provides a framework for DNA replication and genetic processes.

Oxidative phosphorylation A metabolic pathway in mitochondria that creates adenosine triphosphate (ATP) from the oxidation of nutrients. Energy is stored in 'high energy' bonds in ATP.

Phagocytosis Literally 'cell eating', where processes that extend from the cell engulf and enclose material which then becomes internalized and contained within a vacuole in the cytoplasm.

Prokaryote A cell lacking an enclosed nucleus; simpler and considerably smaller than a eukaryote. All bacteria are prokaryotes.

Transcription The synthesis of an RNA copy of DNA in the nucleus as the first step towards protein production in the cytoplasm.

Translation The synthesis of proteins from an mRNA template produced by ribosomes in the cytoplasm.

Zygote The result of fusion between gametes, restoring the diploid cell, and the ultimate stem cell from which the whole organism develops.

Index

A

B

C

D

E

F

G

H

I

J

K

L

M

N

O

P

Q

R

S

T

EVOLUTION

A Very Short Introduction

Brian Charlesworth and Deborah Charlesworth

In less than 500 years the relentless application of the scientific method of inference from experiment and observation, without reference to religious or governmental authority has completely transformed our view of our origins and relation to the universe.

This book is about the crucial role of evolutionary biology in transforming our view of human origins and relation to the universe, and the impact of this idea on traditional philosophy and religion. The purpose of this book is to introduce the general reader to some of the most important basic findings, concepts, and procedures of evolutionary biology, as it has developed since the first publications of Darwin and Wallace on the subject, over 140 years ago. Evolution provides a unifying set of principals for the whole of biology; it also illuminates the relation of human beings to the universe and each other.

www.oup.com/vsi

Epidemiology

A Very Short Introduction

Rodolfo Saracci

Epidemiology has had an impact on many areas of medicine; from discovering the relationship between tobacco smoking and lung cancer, to the origin and spread of new epidemics. However, it is often poorly understood, largely due to misrepresentations in the media. In this *Very Short Introduction* Rodolfo Saracci dispels some of the myths surrounding the study of epidemiology. He provides a general explanation of the principles behind clinical trials, and explains the nature of basic statistics concerning disease. He also looks at the ethical and political issues related to obtaining and using information concerning patients, and trials involving placebos.

www.oup.com/vsi

Online Catalogue

Very Short Introductions

Our online catalogue is designed to make it easy to find your ideal Very Short Introduction. View the entire collection by subject area, watch author videos, read sample chapters, and download reading guides.